Disaster Volunteers

Disaster Volunteers

Recruiting and Managing People Who Want to Help

Brenda D. Phillips
Indiana University South Bend, Indiana, USA

Butterworth-Heinemann is an imprint of Elsevier
The Boulevard, Langford Lane, Kidlington, Oxford OX5 1GB, United Kingdom
50 Hampshire Street, 5th Floor, Cambridge, MA 02139, United States

British Library Cataloguing-in-Publication Data
A catalogue record for this book is available from the British Library

Library of Congress Cataloging-in-Publication Data
A catalog record for this book is available from the Library of Congress

ISBN: 978-0-12-813846-5

For Information on all Butterworth-Heinemann publications
visit our website at https://www.elsevier.com/books-and-journals

Publisher: Candice Janco
Acquisition Editor: Brian Romer
Editorial Project Manager: Lindsay Lawrence
Production Project Manager: Sruthi Satheesh
Cover Designer: Christian Bilbow

Typeset by MPS Limited, Chennai, India

Contents

Biography

Brenda D. Phillips, Ph.D., is the Dean of Liberal Arts and Sciences and Professor of Sociology at Indiana University South Bend. Previously, she taught emergency management at Oklahoma State University and has served as a subject-matter expert, consultant, and volunteer for multiple agencies, communities, educational institutions, and voluntary organizations. She is the author of *Disaster Recovery*, *Mennonite Disaster Service*, and *Qualitative Disaster Research*, the co-author of *Introduction to Emergency Management*, and the co-editor of *Social Vulnerability to Disaster* and co-author of the forthcoming *Business Continuity Planning*. She has written numerous peer-reviewed journal articles in the discipline of emergency management and disaster science with direct experience in researching hurricanes, tornadoes, earthquakes, tsunamis, and hazardous materials accidents, much of which has been funded by the National Science Foundation. Dr. Phillips has been invited to assist or speak at disaster programs in the United States, Canada, Mexico, People's Republic of China, Costa Rica, New Zealand, Germany, Sweden, and Australia where she has promoted evidence-based best practices for community safety.

Dr. Phillips firmly believes in the extension of faculty expertise through volunteer service.

With over thirty years of experience in the field of emergency management education, Dr. Phillips has volunteered for local emergency management planning committees and voluntary organizations, especially for high risk populations. She has served as an unpaid reviewer of city and agency emergency management plans and assisted with planning around disaster-time domestic violence, safety for people with disabilities, and elderly evacuation. She has led business continuity planning at multiple academic institutions and businesses. Her most meaningful volunteer activities have helped to rescue animals and rebuild homes (and docks that support sustainable livelihoods) after disasters.

Foreword

It is a professional and personal honor for me to introduce this much needed, profound contribution to the field of emergency management and beyond that to the whole government, nonprofit, and private sectors which play vital roles in the broadest array of human activity, especially in serving the needs of our communities and their richly diverse populations, in their greatest hours of need.

Dr. Brenda D. Phillips is an amazing and inspiring author and I am pleased to have already read three of her books. I am beyond delighted to see her take on one of the most challenging aspects of disaster management, that of the history, literature, challenges, and opportunities with volunteers. This book is a perfect extension of one of her recent books, *Disaster Recovery*, which I use, as a textbook, in my graduate course on Human Considerations in Disasters at the University of Nevada at Las Vegas, along with Jose' Andres' book, *We Fed An Island*, about his use of volunteers to ramp up the feeding operation in Puerto Rico in 2017 after it was hit by multiple hurricanes.

Dr. Phillips leads the field in expertise and experience in this challenging area of preparedness, mitigation, resilience, response, and recovery, especially in higher education, training, research, and practice. Her approach is very inclusive, interdisciplinary, intergovernmental, interagency, and builds on the amazing capacity of the human spirit to survive, thrive, and to serve others.

Government programs have been designed to promote volunteering and serving in disasters – the Corporation for National and Community Service and the Community Emergency Response Teams (CERT), through FEMA and local emergency management.

Institutions of Higher Education offer related courses and this book is a perfect anchor for any such programs and organizations working in the field. Success in managing Disaster Volunteers can make the difference between creating a successful response and recovery and creating "a disaster within a disaster." This book provides us with the ammunition and guidance to achieve the former and avoid the latter.

Kay C. Goss, CEM®

Professor and Practitioner of Emergency Management,
Former Governor's Senior Advisor for Emergency Management, Fire Service,
Emergency Medical Services, and Public Safety, and Presidential Appointee,
Confirmed by US Senate, as Associate FEMA Director

Acknowledgments and Dedication

I would like to thank my team at Elsevier Press for making this book possible. They include Lindsay Lawrence, Katerina Zaliva, Brian Romer, Ashwathi P. Aravindakshan, Sruthi Satheesh and Candice Janco. Truly, no book ever appears without a hard-working editorial team behind the effort. I am grateful to their dedicated help and guidance. I also thank my dear friend, Sam Elkins, who had my back through the writing process. Her encouragement is deeply appreciated.

As a researcher and author, I gravitate toward topics on which we do not know much scientifically. My aim is to gather what we do know, glean practical lessons from that content, support it with examples and research-driven insights, and inspire both scholars and practitioners to use and expand on what we do know. I spent a considerable amount of time trying to unearth useful pieces on volunteering in the context on disasters. Yet, that body of knowledge still requires a fair amount of research, a comment that I hope inspires the next generation of disaster scientists. Toward that end, this book organizes knowledge into chapters that should surface both useful insights and questions for future research. The last chapter introduces necessary kinds of research to generate even more evidence-based best practices.

We need to get people back home after disasters. This book is dedicated toward that effort. The chapters also offer a walk through how to manage volunteers from the perspectives of emergency managers and volunteer coordinators. A number of lessons can be found for such practitioners and leaders, as well as for students who seek to learn more about recruiting, training, and managing disaster volunteers.

I also hope the volume inspires volunteers to do the right thing at the right time in the right places. We need their help, mostly in the long-term recovery process that can linger for years. However, many volunteers want to help during the immediate aftermath – my hope is that this work inspires them to return for the long haul.

I close by dedicating this work to my beloved Graycie, a blind toy poodle who changed our family forever when we rescued her from a life of terror

PHOTO A1
Graycie tucked her baby under a blanket during the polar vortex. Remember, pets, service animals, and livestock need volunteers too.

and abuse. She stayed by my side during the effort to create this book, and kept reminding me to mention that animals need volunteers too. When you read those examples, remember who inspired them. Rest in peace Graycie (see Photo A1), and know that you made a difference.

CHAPTER 1

Why do people volunteer?

Introduction

Hurricane Katrina, which impacted the U.S. Gulf Coast in August 2005, challenged the volunteer community for well over a decade. Entering with Mennonite Disaster Service as a volunteer, I soon became caught up in the voluntary agency "family," a term I use with intent. Disaster volunteer work often feels like a family reunion with warm embraces between people who haven't seen each other since the "last one." They catch up as they settle into the new work, asking each other how they have been "since Sandy...Haiti.... that wildfire..." and other previously shared projects. Sometimes they inquire about a client they helped together on the last one: "how *is* Maurice? Did the roof get on? Has he moved back in yet?"

They *care*: about the work, the people they help, and each other. They care – because volunteering is how they have been raised, in a socialization process that has made volunteering become part of who they are and what they do. Together, they face the destruction and discern places, tasks, and people in need. They look for meaningful work, set up operations, and start to help. Volunteers lift the burdens of those harmed by disaster and become valuable resources to communities, emergency managers, and volunteer coordinators.

Lifting the burden is exactly what happened after hurricane Katrina, when thousands of volunteers and hundreds of volunteer organizations came from across the U.S., Canada, and around the world to help people in the bayous, coastal villages, and urban neighborhoods along the U.S. Gulf Coast. For the volunteer organizations and emergency managers who tried to coordinate volunteers, the task of where to start felt immense and overwhelming, especially given limited access into flooded areas. So much need existed in an area the size of the United Kingdom. Knowing voluntary organizations could only do so much became a staggering weight to sort through: Who should go where and do what? Which people or places should get help first? Who might not be helped? There was just so much to do "this time."

Disaster Volunteers. DOI: https://doi.org/10.1016/B978-0-12-813846-5.00001-0

And volunteers wanted to come – not understanding they could not yet get in to the badly damaged areas. People jammed phone lines of voluntary organizations, drove themselves to inundated communities, planned mission teams, loaded vans with donations – and created the biggest multi-year volunteer disaster turnout in U.S. history. Managing, harnessing, and leveraging their skills and energy would prove to be an immense task.

Why did so many people volunteer? What motivates people to leave their work, home, school, profession, or family to care for complete strangers? How can receiving communities manage such an influx? How will arriving volunteers be managed and coordinated? Equally important, what do volunteers and aid recipients alike experience from such contributions?

This volume sorts through the process of how to help after a disaster. In this chapter, we explore how disaster volunteering has evolved, why people volunteer, and the kinds of benefits that accrue to survivors, communities, and volunteers. Subsequent chapters will reveal strategies for working with volunteers through evidence-based best practices. This volume aims to support emergency managers and voluntary organizations as they leverage the enthusiasm and energies of eager volunteers. Volunteers will also benefit from this volume as they learn how and where to best fit in and how to make a difference. First, we will look at how disaster volunteering has evolved over time.

A brief history of disaster volunteering

People spend months away from work, school, and family to rebuild the homes of complete strangers. As volunteers, they clear debris, serve food, care for lost pets, put on roofs, provide medical care, use vacation time, travel long distances, spend their own money, and endure harsh conditions in disaster-strewn locales. From spontaneous help to organized efforts, volunteers always show up after disasters. We can count on them, because people have always volunteered after calamity strikes. In this section, we look briefly at examples of volunteering in disaster coupled with how events have driven both numbers and forms of volunteerism.

Historically, volunteer efforts extend back into time and around the world. The Johnstown, Pennsylvania (U.S.) flood of 1889 represents the first documented disaster volunteer effort although other accounts identify Noah as the first to undertake such work (Dynes, 2003; FEMA, 1999; McCullough, 1987). Assisting thousands of Johnstown flash flood survivors, Red Cross volunteers offered mass care including food, first aid, shelter, and hope. Eleven years later, six thousand Galveston, Texas residents perished in a catastrophic hurricane that splintered the U.S. coast (Larson, 2000). The Red Cross came again, along with the Salvation Army. In 1906, a 7.8 Richter magnitude earthquake shook San Francisco, causing building collapses and widespread fires. Dozens of volunteer organizations arrived including the Red Cross, the Salvation Army, the Volunteers of America, and the U.S. Army (FEMA, 1999). Time would prove that an increasingly heavy volunteer turnout arrived after every disaster, laying a foundation to count on, and a need to coordinate, such efforts.

In looking back, it is clear that large-scale disasters have altered how voluntary organizations manage volunteer efforts. After hurricane Camille damaged the U.S. Gulf Coast in 1969, arriving organizations created the National Voluntary Organizations Active in Disaster (NVOAD) in the U.S. to coordinate turnout. The NVOAD movement spread to include state and local chapters, where voluntary organizations could train, network, and use resources wisely. Voluntary organizations began to specialize in specific areas, such as the United Methodist Committee on Relief (UMCOR) that focuses on case management or the Seventh Day Adventists who support donations warehouses. Such efforts leverage volunteer turnout and provide a solid basis for a more efficient distribution of service efforts. In 1992, hurricane Andrew damaged southern Florida which resulted in the United States' using its new response plan. Under Emergency Support Function #6 (Mass Care), NVOAD members formally participated in the plan. To this day, a federal employee named the Voluntary Agency Liaison (VAL) coordinates between governmental and non-governmental sectors under ESF #6. The VAL provides resources and information, meeting space and communications, and other needs for volunteer organizations arriving to help.

Similarly, the 1995 Kobe earthquake prompted Japanese citizens to transform a culture where volunteerism arose out of mutual obligation rather than altruistic foundations. The Great Hanshin-Awaji (Kobe) earthquake resulted in the deaths of over 6000 people with significant damage or destruction to well over 250,000 homes (Kako & Ikeda, 2009). A need to help pushed forward an emerging practice of "borantia" or volunteerism given out of free well (Georgeou, 2006). Such a change in how people viewed volunteerism as a service toward others rather than a duty or mutual obligation under previous authoritarian governance. This cultural transformation resulted in a significant outpouring of aid when the 2011 Tohoku earthquake, tsunami, and nuclear plant disasters occurred. Over one-third reported that they served as a means to "pay it forward." People who had been helped in previous disasters engaged in volunteerism without expectation of a reward as a reflection of a new Japanese way of thinking and behaving (Daimon & Atsumi, 2018).

Australian bushfires, which have devastated large areas of the country, have also prompted volunteerism, from formal to informal efforts. Fire brigades composed entirely of volunteers have formed to battle bushfires in several areas (Webber & Jones, 2011). The 2009 Black Saturday bushfires in Victoria resulted in informal local volunteers providing food and water for people and animals as well as dealing with debris and restoring infrastructure. More formally, people have volunteered through existing organizations like the St. Vincent de Paul Society or Catholic Family Services. As we have come to expect in disasters, people also formed new, local recovery committees to overcome the disaster and to lay a foundation for future emergency responses (Webber & Jones, 2011).

Indeed, when volunteers and voluntary organizations do not identify, understand, or meet all needs, new emergent groups or organizations may form around unmet needs (Enarson & Morrow, 1998; Parr, 1970; Rodriguez, Trainor, & Quarantelli, 2006; Stallings & Quarantelli, 1985). After disasters in Haiti (Farmer, 2011), Pakistan (Sayeed, 2005) and Nepal (Sanderson & Ramalingam, 2015), such emergent groups formed to protect children and women at risk for human trafficking. A well-known problem, the situation often remains under-recognized until it is too late and people have disappeared. As another example of emergence, residents in Watsonville, California organized themselves into a committee to secure more information and resources for Spanish-speaking earthquake survivors after the 1989 Loma Prieta earthquake. Women affected by Hurricane Andrew in 1992 created a Women Will Rebuild organization to work toward the needs of women and children living in Florida (Enarson & Morrow, 1998).

In addition to revealing unmet needs, major disasters have also prompted attention to *how* we volunteer, especially in international settings. The 2010 Haiti earthquake, where an estimated 200,000 + perished, brought in international rescue and relief teams from around the world. The earthquake severely damaged Haiti's capital, Port-au-Prince. Transportation arteries, the water and air ports, and the city's infrastructure sustained significant damage. Hospitals and clinics lay in ruins, and over a million people made their way into relief encampments lacking security, food, water, and hygiene. As teams attempted to enter the area, they found their way blocked by debris which threatened abilities to provide even basic levels of life-saving care. Volunteers found themselves working through extraordinarily difficult conditions. The inundation of volunteers also brought critiques, including some who described the help as "disaster tourism" that provided more photo opportunities than actual aid (Van Hoving, Wallis, Docrat, & De Vries, 2010).

Something similar occurred after the 2004 Indian Ocean tsunami, when volunteer organizations and individuals sent boats to Indian fishing villages affected by the inundation. Unfortunately, the area fishing industry needed different kinds of boats than were donated in order to sustain local livelihoods. As the recovery continued, well-meaning organizations, donors, and volunteers rebuilt housing in India without much consideration from local people or environmental conditions. As a result, rebuilt homes became sweltering locales that provided little relief. Called the donor-driven approach, studies report that such reconstruction often results in under-use because relief organizations do not involve locals in recommending what would work for their climate, livelihoods, and cultures (Karunasena & Rameezdeen, 2010). The lesson learned in this disaster is that local people must be involved in volunteer efforts that influence their lives and well-being both in the immediate aftermath and well into the recovery.

Several lessons thus emerge from this brief overview. First, people have and will always volunteer in disasters and often in massive numbers. We can count on their willingness and enthusiasm and should be ready to manage their arrival. The second lesson is that while people are well-meaning, they may also misunderstand the local context, overlook critical needs, or unintentionally cause problems. In short, we need to be smart about volunteering, both as individuals who serve and as emergency/volunteer managers.

Why people volunteer

Why do so many people want to help? People volunteer because we raise them to do so. By instilling our values in our children, we foster behaviors that benefit society (Amato, Ho, & Partridge, 1984). Sociologists call this process socialization, which is accomplished by agents of socialization including family, teachers, peers, civil society, and faith communities. Generally, the most powerful agent of socialization is the family. With disaster volunteers, both family and faith turn out to be highly influential as people become oriented to assist others.

As the key agent of socialization, families raise us to care for young and old, to complete daily tasks such as preparing meals and cleaning homes, and to secure income that supports our families. We are taught, at a young age, to play well with others, to share our toys, and to be helpful. By socializing family members to be pro-social, we learn to value taking care of each other, especially when they are in distress. Such an early learning experience sets us out on a lifetime journey of engaging in behaviors mutually beneficial to each other and to those around us.

Faith communities also socialize people to be helpful and to care for others (Becker & Dhingra, 2001). For example, faith traditions provide consistent messages to care for those less fortunate, from local soup kitchens to international mission teams. Indeed, we are more likely to volunteer when someone from our faith network asks us to do so (Nelson & Dynes, 1976; Park &

Smith, 2000). Not surprisingly, a majority of disaster volunteer organizations come out of the faith-based community (Nelson & Dynes, 1976; Ross, 1974; Ross & Smith, 1974; Smith, 1978). All faith traditions have inspired organizations to be helpful during disasters. For example, Buddhists formed the Tzu Chi organization to assess disaster impacts and provide funds to survivors. Catholic Charities offers financial aid for rebuilding homes as well as health care, often addressing unmet needs. Protestant traditions also organize disaster response teams across denominations, such as Mennonite Disaster Service which contributes to repairs, rebuilding, casework, and volunteer team management. Within Jewish communities, the Israeli Network for International Disaster Relief offers humanitarian aid, trauma care, and health support. ICNA Relief, a Muslim effort, rebuilds communities and strengthens families with food, shelter, emergency funds, medical, emotional and spiritual care, and case management.

Other agents of socialization, such as political systems, socialize children into beliefs and behaviors that promote civic-minded service. Schools and colleges, also important agents of socialization, teach students to be helpful to each other through service activities. For adults, civic organizations compel people to volunteer. To illustrate, Rotary International partners with ShelterBox to protect disaster survivors from exposure to the elements. Lions Clubs have sent mobile eye clinics to help people replace lost eyeglasses.

Essentially, agents of socialization teach people to behave altruistically, defined as an intentional act to serve others, even to the point of significant self-sacrifice (Piliavin & Charng, 1990). When Hurricane Harvey damaged Houston, Texas (USA) in 2017, two volunteers involved in waterborne rescues died trying to save others. The terror attacks of September 11, 2001 also brought out thousands of volunteers despite potential dangers from airborne particulates. Years later, a health registry has documented lingering and even fatal physical health conditions, particularly among first responders and those involved in clean-up efforts (Friedman et al., 2011).

Disasters rarely occur in comfortable locations either, including hot and humid conditions, isolated locations, or difficult situations. The 2004 Indian Ocean tsunami impacted over a dozen nations. Volunteers traveled significant distances to help, going into or near areas beset with long-standing conflict, such as Sri Lanka. Hurricane Maria devastated multiple islands in 2017, including Puerto Rico. Months went by before power companies restored electricity, creating acutely uncomfortable and sometimes life-threatening conditions for both survivors and incoming volunteers.

Still, people come when disaster strikes. One study found that 10% of all Americans volunteered in some way after September 11, largely because of high levels of identification with those affected (Beyerlein & Skink, 2008).

After September 11, "most expressed that the need to volunteer was overpowering" (Lowe & Fothergill, 2003, p. 298). Blood donors overwhelmed collection sites in the days following the terror attacks, and did so again after the mass shooting in Las Vegas sixteen years later. People raised and donated money for wide-ranging needs from funeral expenses to extended physical therapy and psychological support in both events.

Such altruistic behaviors pour out for multiple reasons (Dynes, 1974). Individual altruism, where we are raised to support others, compels people to give after disaster strikes, such as money, blood, or time. And we also give collectively, when as taxpayers we contribute to programs that help people who have sustained losses or face significant challenges such as aging or homelessness. For disasters, situational altruism arises as we believe that the images we see mean that people's needs are going unmet and we must act (Dynes, 1974). Overall, the various kinds of altruism – individual, collective, and situational – mean that a significant amount of human capital stands ready to make a difference even in the worst conditions and times. What volunteers generate forms a foundation enabling people and communities to arise resiliently from tragedy. Interestingly, volunteers also accrue benefits from their actions.

Benefits of volunteering

Volunteers generate considerable benefits both to themselves and to the broader society, forming social capital akin to a financial investment. Defined as "networks and norms that facilitate collective action" (Woolcock, 1998), social capital appears in the form of free labor, expertise, and goodwill – and can be leveraged. In some contexts, social capital can be valuated as an in-kind contribution for such leveraging. For example, Joplin, Missouri (USA) calculated the value of over 102,000 tornado volunteers at $8.5 million USD. To secure federal dollars, Joplin used the valuation as their 25% required federal match (Abramson & Culp, 2013; Seeley, 2014).

Volunteer managers should recognize and apply several kinds of social capital (Nakagawa & Shaw, 2004; Woolcock, 2002). Bonding social capital (see Table 1.1) defines volunteers with similar interests, skills, and expertise such as spiritual care or construction. Their shared interests create a basis for communities to address mass needs. After the 2004 Indian Ocean tsunami, faith leaders (who hold generally similar backgrounds and outlooks despite denominational specifics) provided spiritual care within and across diverse belief systems. Bonding social capital also comes from civic organizations such as Rotary, Lions, Elks, Moose and others with beliefs that support public service. The Naperville, Illinois, Rotary paid for dozens of homes to be

Table 1.1 Types of social capital.

Type of social capital	Definition	Example
Bonding	Develops out of similar interests, skills	Connections from those active within spiritual care, warehouse expertise, disaster child care
Linking	Arises between interest groups or organizations	Collaborations between government and non-government; fundraising collectively
Structural	Comes from someone's status	Elected official, homemaker, janitor, banker, athlete, small business owner who volunteer their insights, resources, and expertise in a disaster
Cognitive	Ways of thinking about volunteering and disaster	Green reconstruction, accessible or universal design, woman-friendly relief efforts
Bridging	Occurs out of differences between interest groups	Linking skills, expertise to rebuild a home that requires electrical, plumbing, roofing, painting, general construction repairs

Based on Woolcock, M. (2002). Social capital in theory and practice. *Washington D.C.: The World Bank. Retrieved from http://poverty.worldbank.org/library/view/12045/; Nakagawa, Y., & Shaw, R. (2004).* Social capital: a missing link to disaster recovery. International Journal of Mass Emergencies and Disasters, 22, *1*, 5–34.

rebuilt in Pass Christian, Mississippi after hurricane Katrina. In contrast, bridging social capital develops across differences. Rebuilding a single home requires varying kinds of interests, skills, and experiences like finance, case management, construction, plumbing, electrical repair, engineering and more. When people need repairs and reconstruction assistance, wise volunteer managers look for people with both bonding and bridging capital.

Communities also rely on linking social capital, a benefit that arises between interest groups or organizations, such as when voluntary organizations collaborate with government (Brudney & Gazley, 2014). Rotary International, for example, funds ShelterBox efforts during humanitarian crises. Both Rotary and ShelterBox link financing to emergency supplies for people in need. In another classic example of linking, the Southern Mutual Help Association (SMHA) in Louisiana (USA) conducted fundraising and offered reconstruction loans after hurricane Katrina. The SMHA funded clients identified by local Long Term Recovery Committees (LTRCs) composed heavily of local volunteers (see Fig. 1.1). The LTRCs then worked with arriving volunteer organizations, the majority of whom represented faith-based organizations (FBOs) experienced in disaster relief from outside the state.

Structural social capital provides additional benefits. Emanating from a person's status in society, or the position they hold like janitor, banker, or pastor, each position carries with it knowledge, insights, and resources that can infuse volunteer efforts. A business owner could bring managerial skills to organize volunteer work crews, for example, while a maintenance worker

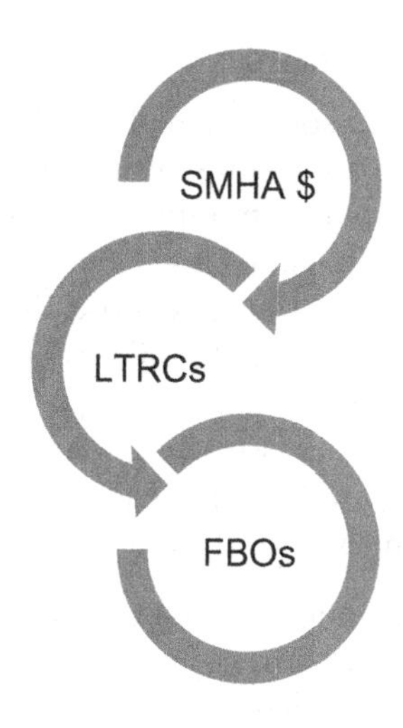

FIGURE 1.1

Linking Social Capital, Hurricane Katrina, Louisiana (USA).
After hurricane Katrina, the Southern Mutual Help Association secured funds to help with recovery. Long Term Recovery Committees then identified clients and conducted case management. Faith based organizations (among others) then sent volunteers to offer repairs and reconstruction assistance. *Source: Phillips, B. (2014).* Mennonite disaster service: building a therapeutic community after the Gulf Coast storms. *Lanham, MD: Lexington Books.*

might understand how things work best under certain conditions. As another kind of social capital, the ways that people think and feel can generate resources. Termed cognitive social capital, viewpoints can make a difference in how communities experience volunteerism. New views in post-disaster housing construction have produced environmentally friendly or "green" reconstruction as well as universal designs for people with accessibility needs. After the 2010 Haiti earthquake, Two Presbyterian synods combined their efforts. Called *Living Waters for the World* and *Solar Under the Sun,* the collaborating synods have placed over thirty solar-powered water treatment systems in or near earthquake-damaged areas. As another example, a Biloxi, Mississippi Lions Club built a camp for children with visual challenges, the only such park on the U.S. Gulf Coast for children with accessibility needs. The transformation of Japanese culture mentioned earlier also represents the power of changing one's outlook.

Identifying kinds and sources of social capital thus emerges as an essential task for volunteer and emergency managers. Leveraging the full range of social capital can enhance volunteer efforts, forge alliances to address disaster needs, alleviate the burdens of people struggling to recover, and accrue both tangible and intangible benefits for communities, volunteers, and recipients.

Community benefits

In addition to the social capital brought by individual volunteers, organizational capital also develops. Faith-based organizations (FBOs) in particular seem to engender community trust, arising from traditions of maximizing resources toward those in need. Such stewardship and use of resources, particularly finances, unburdens those in need and reassures donors. Without such help, people with disabilities, senior citizens, marginalized minority populations, the poor, and single parents would face prolonged routes back to their communities. Knowing that one's insurance money or government aid will be well-used by trusted organizations relieves those beleaguered by recovery challenges (Phillips, 2014).

Volunteer organizations also organize themselves less bureaucratically, meaning they can often address unmet community needs more flexibly than government (Phillips & Jenkins, 2009). That might mean personalizing a home for disability access or rebuilding a home in a remote location that other contractors will not take on. Voluntary organizations also offer specific services usually unavailable from government: day care, roofing, warehousing supplies, mobile showers, case management, donations management, and spiritual care (Bell, 2001; Earl, 2007; Stough, Sharp, Decker, & Wilker, 2010). A good example is "Camp Noah" offered by Lutheran Social Service of Minnesota (USA) where children interact with trained counselors in therapeutic settings (see www.lssmn.org/camp_noah, last accessed July 22, 2015). After the 2001 Gujarat, India, earthquake, grass roots organizations uncovered needs for child care, health facilities, training, and work within a context where gender and poverty intersected to marginalize women (Enarson, 2001). Grass roots organizations offered resources, training, and loans, enabling women to care for themselves and their children. By offering such flexible and focused help, organizations amass and distribute benefits to those in need.

Voluntary organizations may also benefit local community organizations impacted by a disaster. Local agencies may be damaged, or may be so overwhelmed with needs that they experience personnel burnout or loss of regular donations. By infusing help into local agencies, the affected community can restore and replenish their capacities. In return, community organizations can help outsiders navigate the complexities of local political, economic, and cultural contexts and provide insights into the area's customs, language, faith, formal and informal leadership, food, and more.

Experienced organizations also link locally-impacted communities to well-established inter-organizational relationships. The example of NVOAD earlier demonstrates how reputable organizations offer sound practices to benefit the disaster-stricken. NVOAD members know each other well after decades of response and share Points of Consensus around volunteerism and related practices (see www.nvoad.org). Such umbrella organizations can introduce and link LTRCs to specific resources such as debris removal volunteers, roofing teams, or disaster child care services (Peek & Sutton, 2008). At the international level, the United Nations cluster system links various kinds of organizations to specific needs like food security, sanitation, nutrition, camp management, and emergency communications. By connecting organizations and their missions to disaster time needs, the units and their volunteers can accomplish more together than separately, suggesting that a form of inter-organizational capital may also accrue.

Another community benefit comes from the volunteer experience itself. Broadly speaking, people who enjoy positive volunteer experiences tend to volunteer again. Called the "spillover effect," volunteerism generates additional service in other areas (Becker & Dhingra, 2001). A positive disaster volunteer experience will then increase the chance of that same person or group of people volunteering again. Women who volunteered for the U.S. civil rights movement in the 1960s embodied the spillover effect, when they continued to serve their communities decades later in other capacities (Adams, 1990). A similar pattern seems to be in play for disasters as well, with volunteers paying it forward after earthquakes in Japan or hurricanes in the U.S. (Daimon & Atsumi, 2018; Phillips, 2014, 2015).

Benefits to volunteers

In addition to motivations produced by altruistic socialization mentioned earlier, people may volunteer because they derive personal and professional benefits. In general, such benefits may include:

- Improved psychological health such as lowered rates of depression and increased happiness.
- Better physical health, which may be even more pronounced for older volunteers.
- Enhanced corporate images of employees who turn out to support their community.
- An item for one's resume.
- Personal or professional development, such as construction work, management abilities, teamwork/collaboration insights, improved communication capabilities, or growth within personal social interactions.
- Heightened awareness of the impacts of disasters.

- Increased sensitivity to others' cultures and places for living and working (Adams, 1990; Amato, Ho, & Partridge, 1984; Becker & Dhingra, 2001; Beyerlein & Sikkink, 2008; Isham, Kolodinsky & Kimberly, 2006; Lowe & Fothergill, 2003; Lum & Lightfoot, 2005; Phillips, 2014, 2015; Thoits & Hewitt, 2001; Musick, Wilson, & Bynum, 2000).

Specific to disasters, one study of hurricane Katrina revealed multiple levels of benefits to volunteers (Phillips, 2014). Survey respondents reported a new awareness of the powerful effects of disasters. Volunteers also came to understand why people stay in dangerous places: because they love their way of life and feel a kinship to the land, environment, and sea. They also developed meaningful ties with the clients on whose homes they worked, seeing that their hard effort in a difficult climate made a difference. Volunteers also felt that their faith had been strengthened and said they felt a deeper connection to other believers. Other benefits included learning about culturally-based food, music, and story-telling as they served those recovering from the catastrophe (Phillips, 2014).

Benefits to recipients

Many organizations tally benefits to recipients through quantifying the numbers of repairs, rebuilds, or volunteers sent. Or, they publish annual reports noting the amount of funding received and how they accounted for donations. Photos of effective volunteers working in disaster conditions illustrate the success of their mission. However, scant evidence reveals how the beneficiaries of disaster volunteerism experience such efforts.

One study examined how survivors experienced disaster volunteerism (Phillips, 2014). During the extended recovery, survivors felt neglected by government and forgotten by the general public. Significant dissension had erupted, with vocal critiques of relief efforts including negativism toward government efforts and people saying they felt abandoned. In instances of conflict or a "corrosive" community, the experience of having well-prepared disaster volunteers serve on their homes produced an unanticipated consequence (Phillips, 2014). Known as the "therapeutic community," the experience of people arriving to help seems to counteract devastation and loss (Barton, 1970). People who had their homes repaired or rebuilt by volunteers said that their faith in humanity returned. A number of beneficiaries subsequently volunteered in their own communities or on other disaster sites, suggesting that a spillover or "pay-it-forward" spirit may result similar to that described in Japan earlier (Phillips, 2014; Phillips, 2015).

Conclusion

People volunteer because we raise and inspire them to do so. When disaster strikes, we can count on people to come forward and make a difference in communities suffering from grief and loss. Managing and leveraging their devotion can make a difference in the quantity of those served as well as the quality of the volunteer effort that is produced. Making sure that volunteers serve effectively, and that they help rather than harm, is the focus of the remaining content. Coming chapters discuss who manages disaster volunteers, from emergency managers to voluntary organizations (Chapter 2) and what to expect from volunteers (Chapter 3). Chapter 4 then addresses how to recruit and train disaster servants, followed by Chapter 5 which offers best practices for managing disaster volunteers. Readers then learn more about the types of volunteers including volunteers who arrive spontaneously (Chapter 6), align with experienced organizations (Chapter 7), and serve internationally (Chapter 8). Chapter 9 then provides additional insights into the benefits of volunteering with Chapter 10 analyzing the future of disaster volunteerism. Throughout, case examples, guidelines, and checklists will be offered to make the task of managing volunteers easier. Thank you for joining this journey about how to use volunteers effectively in the service of humanity.

References

Abramson, D., & Culp, D. (2013). *At the crossroads of long-term recovery: Joplin, Missouri six months after the May 22, 2011 tornado*. New York: National Center for Disaster Preparedness, Columbia University.

Adams, D. (1990). *Freedom summer*. New York: Oxford University Press.

Amato, P., Ho, R., & Partridge, S. (1984). Responsibility attribution and helping behaviour in the Ash Wednesday bushfires. *Australian Journal of Psychology, 36*(2), 191–203.

Barton, A. (1970). *Communities in Disaster*. Garden City, NY: Doubleday and Company.

Becker, P., & Dhingra, P. (2001). Religious involvement and volunteering: implications for civil society. *Sociology of Religion, 62*(3), 315–335.

Bell, H. (2001). Case management with displaced survivors of hurricane Katrina. *Journal of Social Service Research, 34*(3), 315–335.

Beyerlein, K., & Skink, D. (2008). Sorrow and solidarity: Why Americans volunteered for 9/11 relief efforts. *Social Problems, 55*(2), 190–215.

Brudney, J., & Gazley, B. (2014). Planning to be prepared: An empirical examination of the role of voluntary organizations in county government emergency planning. *Public Performance and Management Review, 32*(3), 372–399.

Daimon, H., & Atsumi, T. (2018). "Pay it forward" and altruistic responses to disasters in Japan: Latent class analysis of support following the 2011 Tohoku earthquake. *Voluntas, 29*(1), 119–132.

Dynes. (2003). Noah and disaster planning: the cultural significance of the flood story. *Journal of Contencies and Crisis Management*, 170–177.

Dynes, R. (1974). *Organized behavior in disaster*. Newark, DE: Disaster Research Center.

Earl, C. (2007). Case management: The key that opens the door to recovery. Retrieved 2012, from http://www.katrinaaidtoday.org/casekey.cfm

Enarson, E., & Morrow, B. H. (1998). Women will rebuild Miami: A case study of feminist response to disaster. In E. Enarson, & B. H. Morrow (Eds.), *The gendered terrain of disaster* (pp. 185–200). Miami, FL: International Hurricane Center.

Enarson, E. 2001. "We want work": Rural women in the Gujarat drought and earthquake. Based on a Quick Response Grant, Natural Hazards Research and Application Center, University of Colorado, Boulder, CO.

Farmer, P. (2011). *Haiti after the earthquake*. NY: Public Affairs.

FEMA. (1999). *The role of voluntary agencies in emergency management*. Washington DC: FEMA.

Friedman, S., et al. (2011). Case-control study of lung function in World Trade Center health registry area residents and workers. *American Journal of Respiratory Critical Care Medicine, 184*, 582–589.

Georgeou, N. (2006). Tense relations: The tradition of Hoshi and emergence of Borantia in Japan. Centre for Asia Pacific Social Transformation Studies. Thesis, University of Wollongong. Available at http://ro.uow.edu/au/theses/528.

Isham, J., Kolodinsky, J., & Kimberly, G. (2006). The effects of volunteering for nonprofit organizations on social capital formation: Evidence from a statewide survey. *Nonprofit and Voluntary Sector Quarterly, 35*(3), 367–383.

Kako, M., & Ikeda, S. (2009). Volunteer experiences in community housing during the Great Hanshin-Awaji earthquake, Japan. *Nursing and Health Sciences*, 357–359.

Karunasena, G., & Rameezdeen, R. (2010). Post-disaster housing reconstruction: Comparative study of donor vs owner-driven approaches. *International Journal of Disaster Resilience in the Built Environment, 1*(2), 173–191.

Larson, E. (2000). *Isaac's storm: A man, a time, and the deadliest hurricane in history*. New York: Vintage.

Lowe, S., & Fothergill, A. (2003). A need to help: Emergent volunteer behavio rafter September 11th. In J. Monday (Ed.), *Beyond september 11Th: An account of post-disaster research* (pp. 293–314). Boulder: Natural Hazards Center.

Lum, T., & Lightfoot, E. (2005). Effects of volunteering on the physical and mental health of older people. *Research on Aging, 27*, 31–55.

McCullough, D. (1987). *The Johnstown flood*. New York: Simon and Schuster.

Musick, M., Wilson, J., & Bynum, W. (2000). Race and formal volunteering. *Social Forces, 78*(4), 1539–1571.

Nakagawa, Y., & Shaw, R. (2004). Social capital: A missing link to disaster recovery. *International Journal of Mass Emergencies and Disasters, 22*(1), 5–34.

Nelson, L., & Dynes, R. (1976). The impact of devotionalism and attendance on ordinary and emergency helping behavior. *Journal for the Scientific Study of Religion, 15*, 47–59.

Park, J., & Smith, C. (2000). To whom omuch has been given: Religious capital and community voluntarism among churchgoing protestants. *Journal for the Scientific Study of Religion, 39*(3), 272–286.

Parr, A. (1970). Organizational response to community crises and group emergence. *American Behavioral Scientist, 13*, 424–427.

Peek, L., & Sutton, J. G. (2008). Caring for children in the aftermath of disaster: The Church of the Brethren Children's Disaster Services Program. *Journal of Traumatic Stress*, 408–421.

Phillips, B. (2014). *Mennonite Disaster Service: Building a therapeutic community after the Gulf Coast storms*. Lanham, MD: Lexington Books.

Phillips, B. (2015). Therapeutic communities in the context of disaster. In A. Collins, S. Jones, B. Manyena, & J. Jayawickrama (Eds.), *Hazards, risks, and disasters in society* (pp. 353–371). Amsterdam: Elsevier.

Rodriguez, H., Trainor, J., & Quarantelli, E. (2006). Rising to the challenges of a catastrophe: The emergent and prosocial behavior following Hurricane Katrina. *Annals of the American Academy of Political and Social Science, 604*, 82–101.

Ross, A. (1974). The emergence of organizational sets in three ecumenical disaster recovery organizations. *Human Relations*, 23–29.

Ross, A., & Smith, S. (1974). *The emergence of an organizational and an organization set: A study of an interfaith disaster recovery group*. Newark, DE: DIsaster Research Center.

Sanderson, D., & Ramalingam, B. (2015). *Nepal earthquake response: Lessons for operational agencies*. London: ALNAP/ODI.

Sayeed, A. (2005). Victims of earthquake and patriarchy: The 2005 Pakistan earthquake. In E. Enarson, & P. Chakrabarti (Eds.), *Women, gender and disaster: Global issues and initiatives* (pp. 142–151). New Delhi, India: Sage.

Seeley, S. (2014). Personal correspondence.

Smith, M. H. (1978). American religious organizations in disaster: A study of congregational response to disaster. *Mass Emergencies, 3*, 133–142.

Stallings, R., & Quarantelli, E. (1985). Emergent citizen groups and emergency management. *Public Administration Review, 45*, 93–100.

Stough, L., Sharp, A., Decker, C., & Wilker, N. (2010). Disaster case management with individuals with disabilities. *Rehabilitation Psychology, 55*(3), 211–220.

Thoits, P., & Hewitt, L. (2001). Volunteer work and well-being. *Journal of Health and Social Behavior, 42*(2), 115–131.

Van Hoving, D., Wallis, L., Docrat, F., & De Vries, S. (2010). Haiti disaster tourism – a medical shame. *Prehospital and Disaster Medicine, 25*(3), 201–202.

Webber, R., & Jones, K. (2011). Rebuilding communities after natural disasters: The 2009 bushfires in Southeastern Australia. *Journal of Social Service Research, 39*(2), 253–268.

Woolcock, M. (2002). *Social capital in theory and practice*. Washington D.C: The World Bank Retrieved from. Available from http://poverty.worldbank.org/library/view/12045/.

Further reading

Farris, A. (2006). Katrina anniversary finds faith-based groups still on the front lines, resilient but fatigued. New Orleans. Retrieved July 15, 2008, from http://www.religionandsocialpolicy.org.

Fujita, A. (2011). Japan earthquake-tsunami spark volunteer boom but system overwhelmed.

CHAPTER 2

Who manages disaster volunteers? Why manage volunteers?

Case examples

Disasters compellingly pull us, as media images spell out the ways in which people are in trouble. However, too many times, people make assumptions about what kind of help is needed. Volunteers then turn out during the immediate response time, thinking that people need the most help right after an event. However, disasters generate a wide range of needs that may take years to overcome meaning that emergency and volunteer managers truly need to be ready to leverage the help potentially coming their way.

Without such volunteer help, many families and communities would not be able to recover, thus stricken communities need volunteer help. Volunteers can be managed in multiple ways, from highly organized and intentional efforts to strategies that evolve on the spot. Managers and their communities will also need to determine what kinds of volunteers should be invited to help – as well as where, when, and how to use them. In this section, readers will learn about different scenarios that reveal the breadth of need for volunteers and why it is important to send forth who is needed at the right times, to the right places, and with the right kinds of preparedness.

Sri Lanka, Indian Ocean Tsunami and beyond (2004–2005)

On December 26, 2004, a 9.0 Richter magnitude earthquake occurred in the Sumatran area of the Indian Ocean in one of the worst earthquakes ever recorded. A powerful, underwater up-thrust of land caused waves to spread out toward multiple nations. Upon nearing land, the tsunami pulled water out from bays, ports, and beaches and then slammed back onto shore with a massive, deadly impact. Waves as high as 30–40 feet picked up people, buildings, vehicles, trees, animals, bridges, and boats, violently pushing inland for several kilometers in coastal areas of Sri Lanka, India, and Thailand. Catastrophic injuries and loss of life occurred, with as many as 300,000 people perishing in thirteen nations. Hundreds of agencies,

Disaster Volunteers. DOI: https://doi.org/10.1016/B978-0-12-813846-5.00002-2

organizations, and governments deployed non-profit units, militaries, and professional rescue teams. But where to start a volunteer effort in such a massive and widespread disaster site? And, what kind of aid would be needed?

Immediate needs for all affected nations included mass burials or cremations as well as repatriation of the remains of foreign visitors, tourists, and business travelers to their home countries. The power of the tsunami waves also produced significant environmental and agricultural damage both at sea and on land that damaged sustainable livelihoods – a challenge that most voluntary organizations had rarely faced. Simultaneously, millions needed mass care including special circumstances generated for those newly orphaned or widowed.

Many errors occurred in sending volunteer aid to affected areas. Well-meaning volunteers who could not speak the local languages arrived, when their airfare alone could have funded the rebuilding of many homes (Mulligan & Shaw, 2011). Well-meaning donors sent the wrong kind of boats to devastated fisheries as well as clothing unsuitable for the local climate. Well-meaning organizations rebuilt homes, without the input of locals, in what became sweltering homes unusable for the local climate and without the requisite areas needed for cooking, sleeping, and working in various industries (Barenstein, 2006). One again an important lesson had to be relearned: to send the right kind of volunteers for the right reasons to the right places.

The tsunami also displaced close to one million people, many of them living in hundreds of relief encampments (Gelkopf, Ryan, Cotton, & Berger, 2008). Within those encampments could be found specific populations that needed specific kinds of volunteer help. To illustrate, the event claimed lives unevenly, and with 46,000 + dead or missing in Sri Lanka alone, many children were left both injured and bereaved. Caring for children requires carefully prepared adults who can manage a wide array of reactions from fear of future events to deep-seated grieving. Given the number of affected cultures and faiths in any single affected nation, such psychological care must be attuned to local contexts, so that such help feels appropriate and familiar and makes a difference (De Silva, 2006; Gillard, & Paton, 1999; Norris, Friedman, & Watson, 2002; Norris, Friedman, Watson, & Byrne, 2002; Schmuck, 2000).

In Sri Lanka, a local organization invited Psychology Beyond Borders to train disaster volunteer workers who would then train thirty teachers. Trainers relied on experience from offering school-based programs for children originally centered on terrorism and adapted their model to fit the tsunami. Practicing a flexible approach, the creators mindfully worked through issues related to culture, gender, and faith to create an appropriate program called ERASE Stress. Students subsequently received a number of interventions that were developmentally appropriate by age like art therapy and culturally-centered such as meditation (Gelkopf et al., 2008). Meeting specific needs in an intentional way represents a best practice and evidence-based way to help.

Joplin, Missouri, May 22, 2011

In 2011, an EF5 tornado devastated Joplin, Missouri (U.S.). The destruction covered an area three-fourths of a mile wide to 22 miles in length (Paul & Stimers, 2012). A record 162 people died, in over 7000 homes and businesses torn apart along a multiple vortex path. Close to 175,000 volunteers arrived in the first six months after the tornado. But two weeks prior to the tornado, the area's Red Cross had established an agreement with Missouri Southern State University to use their facilities for emergency shelter and volunteer management (Abramson & Culp, 2013). A group called Americorps soon arrived to manage the volunteer turnout descending on the stricken area. In addition, area faith communities drew on their networks to attract volunteers and provide accommodations. Local businesses and organizations, such as realtors and the area's Chamber of Commerce, donated funds for tools (Abramson & Culp, 2013).

By December, in a town of just over 50,000 people, an estimated 114,000 volunteers had been provided with work, housing, and food. An additional 42,000 to 57,000 spontaneous volunteers had gone straight into neighborhoods to help (Abramson & Culp, 2013). Volunteers engaged in traditional disaster tasks as they cleared the streets, sorted and carried debris to the curb for disposal, repaired and rebuilt homes, provided hugs and health care, raised money for school events, and created a permanent memorial to those who died. They helped to re-open the schools and the downtown, conducted fundraisers, and raised the spirits of a community reeling from a terrible night. These kinds of activities represent what many volunteers expect to do in a disaster and usually perform very well with limited supervision. Still, being ready to manage thousands of people made it possible in this more traditional type of disaster event.

Sweden and Australia, Fires, 2009–2014

Increasingly, wildfires or bushfires threaten human populations due in part to climate change amid the places where we build homes and businesses. Due to an overall expectation that such wildfires will continue worldwide, volunteer effort may be crucial to enlist. As an example, a heatwave, high winds, and dry conditions caused multiple fire sites to break out across the state of Victoria, Australia in 2009, resulting in one of the nation's worst-ever disasters (Cameron et al., 2009). This 'Black Saturday' bushfire claimed 180 lives and left many without homes or places of employment. Volunteers do help in such an event. In Australia, close to 25,000 Australians support 750 volunteer fire brigades although recently such volunteer firefighter numbers have been declining (McLennan & Birch, 2005). To address the lack of volunteers, efforts have intentionally focused on increasing women's participation (Branch-Smith & Pooley, 2010).

Another massive fire, the largest in Sweden's modern history, broke out in a forested province called Västmanland in 2014 (Johansson, Danielsson, Kvarnlöf, Eriksson, & Karlsson, 2018). The first volunteers from nearby farming sectors extended their equipment to assist firefighting operations. A few days later, Swedish municipalities tapped into pre-trained Voluntary Resource Groups for additional help (Johansson et al., 2018). As volunteers, people will often show up as organizationally affiliated and trained support that can result in a focused effort. Although spontaneous volunteers did show up, Sweden relied on affiliated and prepared organizations in order to integrate volunteers effectively into difficult firefighting operations (Johansson et al., 2018).

Terrorism, 2001 to present day

People desperately wanted to volunteer after the terrorist attacks of September 11th, 2001 in the U.S. However, terrorism presents an different kind of disaster for volunteers. This time, survivors required financial support and professional care for unusual and massive needs: funerals, burn treatments, psychological recovery, and health care that would span decades. Workers also needed help dealing with lost income, the costs of travel to reunite family, and job searching. The diversity of people in New York City alone, and the visitors to the World Trade Center, meant that dozens of languages needed to be accommodated. Adjacent offices had to be cleaned, but with appropriate protective clothing and resources due to the particulates that contaminated surfaces and the air (Lippman, Cohen, & Chen, 2015).

In other 9/11 sites, such as the Pentagon, volunteering would prove difficult due to the sensitive location of this military center. Imagine a similar attack on a country far from your own, such as the 1998 attack on the U.S. embassy in Kenya (Bushnell, 2018). In addition to supporting Americans affected, the bombing killed and injured numerous Kenyans, compromised security for both nations, and proved difficult to support from a distance. Nonetheless, people do try to volunteer support to those affected. A 2013 terrorist attack on Nairobi's Westgate Mall in Kenya quickly generated volunteers supporting food, water, spiritual, and emotional care during the extended assault. Efforts then transitioned to fund medical and funeral costs. September 11 sites in the U.S. have since been memorialized through plaques, artwork, exhibits, and museums similar to the Oklahoma City National Memorial and

Museum that followed this domestic terrorist attack in 1995. But still, additional impacts proved difficult to support through volunteerism such as addressing a range of issues from lost wages and jobs in the tourist industry to the extended psychological support that traumatized survivors would need (Buigot & Amendah, 2016).

Although people have always helped each other in terror attacks, a 2017 situation in Las Vegas, Nevada (U.S.) revealed the extent of volunteer courage and self-sacrifice. When a shooter opened fire from a high vantage point, 58 concert-goers below died and hundreds more sustained injuries. Yet, even as the attack continued, people helped each other, covered others with their own bodies, pulled the injured to safety, provided first aid, and transported survivors to hospitals. They did not stop there, as an international outpouring of aid crowdsourced funds to help with health care costs and rehabilitation, funerals, travel costs, and lost wages. Similar efforts to save friends and strangers unfolded in Dayton, Ohio in 2019 after a mass shooting followed by fundraising to help those affected. People want to help when crisis strikes.

An attack in Orlando, Florida (U.S., in 2016) claimed lives as well, but this time the terrorist attack targeted the LGBTQ community. New Zealand experienced a similar attack on a Muslim mosque followed by the targeted, mass murders of many Hispanic Americans and Mexican citizens in El Paso, Texas in 2019. In addition to deaths and horrific injuries, harm arose from these intentional acts: hate crimes. In compassionate outpourings, spontaneous volunteers created impromptu memorials, raised funds for funerals and health care costs, and publicly decried the hateful acts. As time goes on, volunteer and professional organizations are likely to provide aid, such as that generated after 9/11 by the Center for Justice and Peacemaking. At this institute, volunteers engage in week-long and multi-level training to work with traumatized people, learn how to facilitate conflict transformation, and participate in restorative justice (https://emu.edu/cjp/star/about). Alumni from the program work to address hate crimes, war impacts, refugee crises, and intergroup conflict worldwide.

Summary

As the previous examples illustrate, disasters generate a wide range of needs (see Box 2.1). Volunteers will be needed and can make a significant difference if managed effectively. With the right approach, volunteers will also have a meaningful experience and perhaps one that transforms their life and perspectives. For those served, the right help will reduce the harm they have suffered and move them toward a restored sense of self and community. This book offers guidance to all, particularly those who manage volunteers' abundant supply of help and hope.

Box 2.1 Kinds of help needed in various kinds of disasters

Indian Ocean Tsunami, 2004.

- Religious support for mourning and grief
- Appropriate rituals for the dead (e.g., funerals, cremations)
- Support newly orphaned children
- Open and support tent cities
- Provide food and water
- Find alternative work
- Restore lost livelihoods
- Address environmental damage
- Stop human trafficking
- Repair and rebuild homes
- Restore hospitals, bridges, roads
- Rebuild and re-open business sectors
- Manage stress and loss

Joplin Tornado, 2011.

- Clear debris
- Rebuild hospital, congregate care facilities, businesses
- Take in lost pets, find owners, host adoption event
- Rescue livestock
- Offer appropriate clothing
- Manage shelters, temporary housing sites
- Repair and rebuild homes
- Help people find new jobs
- Disseminate information in multiple languages
- Manage massive influx of spontaneous volunteers
- Get schools open again, fund events to return normalcy

Wildfires, Sweden & Australia

- Fight fires
- Remove debris including charred homes
- Retrieve anything salvageable
- Address job loss and displacement
- Rebuild homes, businesses, schools, agencies
- Find lost pets, treat injured animals, provide foster care
- Offer emergency medical services
- Support injured people during recovery
- Support grief-stricken

Terrorism

- Protect yourself and others
- Offer first aid
- Donate blood if needed
- Fundraise health care for significant injuries and rehabilitation
- Support people with new, long term disabilities
- Manage conflict
- Address hate crimes
- Provide information and help in many languages
- Assist people visiting from outside affected nation
- Provide spiritual and emotional care
- Support long term psychological care
- Raise funds for unmet, long-term needs
- Participate in public memorials and related events
- Create memorial sites
- Hold annual remembrances

Who manages disaster volunteers

Disaster volunteers arrive in a variety of ways and end up becoming organized and active similarly. They may come to the scene uninvited but eager to help (Sauer, Catlett, Tosatto, & Kirsch, 2014; Twigg & Mosel, 2017) or come with an experienced unit. Regardless, they will come – which, ultimately, can be good news if communities affected by disaster have made themselves ready for the energy headed their way. In this section, we look at a continuum of ways in which disaster volunteers arrive, from spontaneous arrivals to the professionals who should be highly involved in managing volunteers. The remainder of the book will lay out and expand on strategies to manage disaster volunteers in order to push that eagerness to help in the best possible directions. People who want to volunteer will also find content useful as they think through how, when, and where they want to provide the best possible help.

Self-managed volunteers

When disaster happens, the first volunteers will likely be family, friends, and strangers who pull us from the rubble, offer first aid, or call for help (Twigg & Mosel, 2017). Because we raise people to care for others in trouble, volunteers engage in extraordinarily selfless acts like the people who put their lives on the line in a terrorist attack. Given more advance time, self-motivated volunteers may simply show up at a disaster, launch their own emergent effort, or join an affiliated organization. The majority, historically, have shown up on their own in a phenomenon called convergence (Fritz & Mathewson, 1957). Such convergence produces a turnout called spontaneous unplanned/unaffiliated volunteers (SUVs). SUVs can produce both benefits and challenges, which will be further discussed in Chapter 6 (e.g., see Kartez & Lindell, 1987; Twigg & Mosel, 2017).

Historically, televised images seem to convince massive numbers of people that they need to head into the disaster zone. Increasingly, electronic and social media tools are crowdsourcing volunteer help (Ludwig, Kotthaus, Reuter, van Dongen, & Pipek, 2017). Such emergent activity, while new in its form, is the same kind of disaster convergence that has happened for a long time. However, volunteer managers need to be familiar with how people spread information and mobilize around a disaster via electronic media, platforms, and apps. By using the same tools, they may be able to crowdsource help – and potentially crowd manage it - in a meaningful way (see more on this in Chapter 6).

From a volunteer perspective, it just makes sense to go and help because of the damage they can see on various media and because people seem to need assistance. Certainly, people and places do need help but when is the best time to arrive and wade in to the debris? That debris can be dangerous, containing a number of hazards generated by the event. Roads may be inaccessible, and communities may not be ready for hundreds or thousands of volunteers to descend. Thus, self-managed volunteers should listen carefully to officials to guide them into the best places and ways to assist as well as when to arrive. One of the most important pieces of advice in this volume will be to wait until asked and to consider waiting until the long-term recovery time period begins which is when communities need the most help.

For those who manage SUVs, it is important to know that volunteers will come even without being asked. Anticipating their eagerness to help, and channeling it effectively, represents important steps in moving a damaged area from response into recovery. The example of Joplin given earlier in this chapter demonstrates such an effort. Though the area identified a volunteer site just weeks before the tornado, local officials and Americorps coordinated and used arriving volunteers extremely well.

Community based

Communities may also opt to organize cadres of volunteers before a disaster happens. In the United States, an effort called Citizen Corps emerged after the terror attack of September 11 (Franke & Simpson, 2004). The national-level effort provided a framework and training protocols for local and state groups and produced several groups of now-ready volunteers including Medical Reserve Corps, Community Emergency Response Teams, Volunteers in Police Service, and Fire Corps. As noted earlier, Australia has created Volunteer Fire Brigades and Sweden hosts Volunteer Resource Groups. Similar groups may be called Community Organizations Active in Disaster (COADs). Another common post-event effort is the Long Term Recovery Committee/Group (LTRC/G) that convenes disaster volunteers and organizations into collaborative networks to help area survivors. At the meetings, case managers present the situations of their clients while disaster organizations volunteer to resolve lingering recovery issues usually around repairing or rebuilding homes.

The biggest challenge with community-based groups tends to be the problem of turnover. People usually volunteer when they have available time and interest, which can vary based on age, work responsibilities, and family needs. Thus, a call for volunteers may produce a handful of available people or potentially more. A related problem comes from sustaining volunteer interest when help is not needed. Communities rarely face non-stop disasters, so keeping people trained and ready to show up when a flood happens every ten to twenty years can make it difficult to retain volunteers. However, by providing other opportunities such as supporting area fairs, sporting events, public gatherings, health drives, and similar efforts it can be possible to provide a bonding experience that builds a disaster ready team. Local agencies, like public health or homeless shelters can also partner with their volunteers and engage them in regular activities to remain connected (Stajura et al., 2012).

Non-governmental and disaster organizations

Community-based efforts differ from non-governmental organizations (NGOs), which are private, intentional entities usually focused on a humanitarian mission (Werker & Ahmed, 2008). NGOs typically engage in fund-raising, often have paid staff, integrate volunteers when needed, and work daily to address needs. They may be disaster-focused, but they may also include disasters as part of their overall work. NGOs may also be location-specific or operate world-wide. International governmental organizations (INGOs) have further spread disaster volunteerism. From the International Red Cross/Red Crescent to the Salvation Army movement, Oxfam, CARE, and Doctors without Borders, hundreds of thousands of people have served communities distant from their own homes and supported strangers through crisis. Most recently, the United Nations has worked with government and INGO sectors to create a "cluster" system that involves an array of NGOs and government efforts focused on specific needs like logistics, shelter, health, and nutrition.

Disaster organizations may also fit the definition of I/NGOs, but dedicate their efforts exclusively to the disaster. Often called disaster affiliated organizations, their mission centers on restoring people and places through volunteer effort. They likely arise out of a larger set of beliefs or connections, such as the numerous faith traditions that have launched disaster relief units. An example would be Nechama, a Jewish organization that centers on response, recovery, and disaster training. The United Methodist Committee on Relief offers similar services alongside a long-time dedication to disaster social work. Civic sources also produce disaster affiliated organizations, like Team Rubicon that arose out U.S. military veterans volunteering for disaster response work. Businesses may also donate resources to create a support organization, like Tool Bank, which covers an unmet need for disaster tools. The benefit of such organizations is that they arrive through an intentional process, are knowledgeable about what to do and how to collaborate, and come in to a community relatively self-contained. Disaster organizations and their affiliated volunteers represent the backbone of long-term disaster recovery, because they are far more likely to volunteer for years than self-managed volunteers who typically stay for hours to days. Chapters 3 and 7 will include more detail on these types of disaster organizations and their volunteers.

Government

Depending on where the disaster occurs, governmental response and recovery efforts may offer the broadest and deepest source of assistance or the least. Financially-stable or well-developed nations usually have the most robust and well-funded programs, though these programs may be time-limited or target specific needs and populations. For example, nations like France and Germany enjoy considerable resources to respond to floods, heat waves, and terror attacks. Nonetheless, governments may have limits on how long they will fund relief efforts and what kinds of programs they will provide. Although it is unusual, governments may refuse entry to outside voluntary organizations, such as when the Nargis cyclone hit Myanmar in 2008. Most of the time, though, government relief seems concentrated on the response time period, repairable physical damage, and immediate injuries. Fewer programs fund long-term rehabilitation from injuries, offer spiritual or psychological support, or address unique needs – with all of the latter usually coming from NGOs, especially in lower income nations. Governments also face continual, non-disaster needs for their resources and attention, so at some point the funding and programs will likely end.

Economically-challenged countries will likely struggle to provide sufficient relief and recovery help to their people. The case of Haiti represents one situation, where pre-existing conditions made it difficult to survive during normal times. When the 2010 earthquake claimed over 200,000 lives, fatalities included government officials, community leaders, emergency personnel, and I/NGO leaders. Residents in and around the nation's capital desperately needed outside help. Aid inundated affected areas, clogging damaged ports, and creating slowdowns along highways crowded by well-meaning volunteers and donations. Yet, Haiti needed specific assistance: debris removal to aid traffic movement, repairs to the harbor and airport, and re-establishing communications. Simultaneously, immediate needs had to be met: trauma and medical care, food and water, shelter, and safety from human trafficking. The recovery period lingered, with residents still in tent cities years after most volunteers had left. One should not discount the resources available in a developing nation, though, particularly the social capital found within the human spirit. Kenyans have demonstrated this spirit repeatedly during multiple natural disasters and terrorist events.

Governments may also engage in advance planning to manage volunteers. In the U.S., the Federal Emergency Management Agency (FEMA) has a well-developed and battle-worn National Response Framework that includes an Emergency Support Function (ESF) tasked with volunteer integration and management. ESF #6 (see Box 2.2), concentrates on mass care and includes a key person called the Voluntary Agency Liaison (VAL) who convenes and

Box 2.2 Emergency Support Function #6 mass care, emergency assistance, housing, and human services annex (January 2008)

The ESF Coordinator for Mass Care is the U.S. Department of Homeland Security and the Federal Emergency Management Agency, which also serve as the Primary Agency for delivery of related services. As part of the National Response Framework, ESF#6 includes numerous support agencies such as:

- National Voluntary Organizations Active in Disaster
- Other voluntary agency and nongovernmental support organizations
- Department of Agriculture
- Department of Defense
- Department of Health and Human Services
- Department of Housing and Urban Development
- Department of the Interior
- Department of Labor
- Department of Transportation
- Department of the Treasury
- Department of Veterans Affairs
- General Services Administration
- U.S. Postal Service
- American Red Cross
- Corporation for National and Community Service

Source: Federal Emergency Management Agency, 2019.

supports volunteer agencies in its Joint Field Office (JFO). The National Disaster Recovery Framework also recognizes the value of disaster affiliated organizations and the collaboration necessary to help communities rebuild. For all current planning frameworks, visit https://www.fema.gov/national-planning-frameworks, last accessed June 18, 2019.

Professionals

Finally, we turn to the professionals who manage volunteers in disaster. For our purposes, this book will focus on two groups: emergency managers and volunteer managers. Though they may come from very different areas, they may very well overlap during a disaster. Emergency managers offer professional experience in preparedness, planning, response, and recovery. Their depth of knowledge, experience and evidence-based practice can provide a context in which volunteer efforts for disasters will be undertaken. Emergency managers know and understand what a disaster area will look like, including the problems that volunteers will encounter when trying to help. They will see and understand the disaster site differently, including the necessary steps that must be taken to move the community from response into recovery. Increasingly, academic programs train emergency managers on using volunteers for both pre- and post-disaster efforts. However, emergency management agencies vary in capacity and resources from those with a full-time volunteer coordinator to those with a part-time employee who manages the entire agency. In many developing nations, it remains impossible to fund emergency management positions which are often covered by police or fire personnel who may also have limited time and resources.

Volunteer coordinators and managers represent another valuable community asset. An area may have a Volunteer Center, where community-wide volunteer efforts are organized and coordinated. That site could become the best resource for volunteers and emergency managers to use in a disaster. Ideally, the emergency management agency and the volunteer center will have cross-trained and conducted volunteer management planning well in advance of a disaster and be ready to handle both SUVs and experienced disaster organization-based volunteers. In future chapters, readers will see examples of how such a coordinated effort unfolds. Volunteers should thus follow the various Internet, platforms, apps, and social media that local emergency management agencies and volunteer centers produce as a way to know when and how to help best.

Why manage disaster volunteers

This book outlines exactly why and how to manage disaster volunteers, which enables communities to harness and apply the energy and good will coming into their location. Communities will want to find ways to do so. Because disasters produce both physical and psychological harm, the arrival of volunteers can speed the healing of those facing the rubble pile and help survivors shoulder the burden of recovery. Volunteers do help and are nearly always needed, for both general and specific tasks. An organized and mindful approach to managing volunteers thus spurs a return to normalcy. For every elected official who has faced the cameras and said "we will rebuild" – volunteers and their organizations represent the crucial resource that will make it so.

Volunteers also take on difficult work that governments and homeowners cannot do or cannot afford. Insurance, government relief funds, and donations rarely cover all needs, especially for economically disadvantaged populations and places. Volunteer labor steps in where money ends, such as adding a wheelchair ramp to the home of someone with a mobility challenge. Volunteers do that work with a willing spirit designed to help their fellow human beings, with little expectation of reward or recognition. They do that work because, as a society and planet connected to the well-being of other human beings, we have raised them to be there when hurtful events occur.

Because volunteers are there for our people, managers and organizations need to be there for them. Managers need to be ready to guide and support the work of volunteers and their organizations. They must be ready before volunteers arrive or be prepared to organize quickly when disaster strikes. Volunteers will need guidance on where to focus their energy and how to do so safely. Thus, managers need to be ready to register, train, and deploy those that arrive. Managers also need to be knowledgeable about voluntary organization capacity and who can do what in order to make the most of the help headed their way.

Managers, and volunteers as well, also need to recognize that while the majority of help shows up during the response period, volunteers will be needed most during the long-term recovery phase. Clearing the streets, disposing of destroyed furniture, or tearing out the interior of a home prior to rebuilding is just the start of the life cycle that disasters generate. It takes time to repair and rebuild homes and lives, so managers and volunteers alike will want to know where they can best use that unpaid and dedicated human spirit.

Tasks for disaster volunteer managers

The remainder of this book walks volunteers, emergency managers, and volunteer coordinators through the key steps of working on a disaster site together. In Chapter 4, readers will learn about the essential tasks of recruiting and training volunteers to repair damaged communities and heal disrupted lives. Chapter 5 then brings various disaster scenes to life and walks readers through how volunteer management needs to occur to be effective and to make a difference. Subsequent chapters then work through the various challenges and benefits of a range of disasters. Chapter 6 will address the pros and cons of spontaneity and how to manage convergence. Next, Chapter 7 explains the value that disaster-organization affiliated volunteers offer and how to use these resource-rich volunteer pools to nourish the response and recovery periods. Due to the particularly challenging aspects of volunteering globally, Chapter 8 will look into best practices for international volunteering. The book then wraps up by discussing the benefits of volunteering (Chapter 9) and the future of such efforts (Chapter 10).

Conclusion

Volunteer management is essential in order to leverage the energies provided by those who want to help others. Whether at a local level, or in cooperation with experienced disaster organizations, or working internationally to deliver response and recovery support, disaster and volunteer managers – as well as the volunteers themselves – need to coordinate and collaborate. Whether you are a volunteer or an emergency or volunteer manager, you are invited to join this meaningful journey, with deep appreciation for the volunteering and work that you are preparing to do on behalf of those in need.

References

Abramson, D., & Culp, D. (2013). *At the crossroads of long-term recovery: Joplin, Missouri Six Months after the May 22, 2011 Tornado*. NY: National Center for Disaster Preparedness, Earth Institute, Columbia University.

Barenstein, J. (2006). Challenges and risks in post-tsunami housing reconstruction in Tamil Nadu. *Humanitarian Exchange, 33*, 39–40.

Branch-Smith, C., & Pooley, J. (2010). Women firefighters' experiences in the Western Australian volunteer bush fire service. *Australian Journal of Emergency Management, 25*(3), 12–18.

Buigot, S., & Amendah, D. (2016). Effect of terrorism on demand for tourism in Kenya. *Tourism Economics, 22*(5), 928–938.

Bushnell, P. (2018). Terrorism, betrayal, and resilience: My story of the 1998 U.S. embassy bombings. Omaha, NE: University of Nebraska Press.

Cameron, P., Mitra, B., Fitzgerald, M., Scheinkestel, C., Stripp, A., Batey, C., … Cleland, H. (2009). Black saturday: The immediate impact of the February 2009 bushfires in Victoria, Australia. *Medical Journal of Australia, 191*(1), 11–16.

De Silva, P. (2006). The tsunami and its aftermath in Sri Lanka: Explorations of a Buddhist perspective. *International Review of Psychiatry, 18*(3), 281–297.

Franke, M., & Simpson, D. (2004). *Community response to hurricane Isabel: An examination of Community Emergency Response Team (CERT) organizations in Virginia*. Boulder, CO: Natural Hazards Center, University of Colorado, Quick Response Research Report 170.

Fritz, C., & Mathewson, J. (1957). *Convergence behavior in disaster: A problem in social control*. Washington DC: National Academy of Sciences/National Research Council, Publication 476.

Gelkopf, M., Ryan, P., Cotton, S. J., & Berger, R. (2008). The impact of train the trainer course for helping tsunami-survivor children on Sri Lankan disaster volunteer workers. *International Journal of Stress Management, 15*(2), 117–135.

Gillard, M., & Paton, D. (1999). Disaster stress following a hurricane: The role of religious differences in the Fijian islands. *Australasian Journal Of Disaster, 2*, 2–9.

Johansson, R., Danielsson, E., Kvarnlöf, L., Eriksson, K., & Karlsson, R. (2018). At the external boundary of a disaster response operation: The dynamics of volunteer inclusion. *Journal of Contingencies and Crisis Management, 26*, 519–529.

Kartez, J., & Lindell, M. (1987). Planning for uncertainty: The case of local disaster planning. *Journal of the American Planning Association, 53*(4), 487–498.

Lippman, M., Cohen, M., & Chen, L. (2015). Health effects of World Trade Center (WTC) dust: An unprecedented disaster with inadequate risk management. *Critical Reviews in Toxicology, 45*(6), 492–530.

Ludwig, J., Kotthaus, C., Reuter, C., van Dongen, S., & Pipek, V. (2017). Situated crowdsourcing during disasters: Managing the tasks of spontaneous volunteers through public displays. *International Journal of Human-Computer Studies, 102*, 103–131.

McLennan, J., & Birch, A. (2005). A potential crisis in wildfire emergency response capability? Australia's volunteer firefighters. *Environmental Hazards, 6*, 101–107.

Mulligan, M., & Shaw, J. (2011). Achievements and weaknesses in post-tsunami reconstruction in Sri Lanka. In P. Karan, & S. Subbiah (Eds.), *The Indian Ocean Tsunami: The global response to a natural disaster* (pp. 237–260). Lexington, KY: University of Kentucky Press.

Norris, F. H., Friedman, M. J., & Watson, P. J. (2002). 60,000 disaster victims speak: Part II. Summary and implications of the disaster mental health research. *Psychiatry, 65*(3), 240–260.

Norris, F. H., Friedman, M. J., Watson, P. J., Byrne, C. M., Diaz, E., & Kaniasty, K. (2002). 60,000 disaster victims speak: Part I. an empirical review of the empirical lterature, 1981–2001. *Psychiatry, 65*(3), 207–239.

Paul, P., & Stimers, M. (2012). Exploring probable reasons for record fatalities: The case of 2011 Joplin, Missouri tornado. *Natural Hazards, 64*(1511), 1526.

Sauer, L., Catlett, C., Tosatto, R., & Kirsch, T. (2014). The utility of and risks associated with the use of spontaneous volunteers in disaster response: A survey. *Disaster Medicine and Public Health Preparedness, 8*(1), 65–69.

Schmuck, H. (2000). An act of Allah: Religious explanations for floods in Bangladesh as survival strategy. *International Journal of Mass Emergencies and Disasters, 18*(1), 85–95.

Stajura, M., Glik, D., Eisenman, D., Prelip, M., Martel, A., & Sammartinova, J. (2012). Perspectives of community- and faith-based organizations about partnering with local health departments for disasters. *International Journal of Environmental Research and Public Health, 9*, 2293–2311.

Twigg, J., & Mosel, I. (2017). Emergent groups and spontaneous volunteers in urban disaster response. *Environment & Urbanization, 29*(2), 443–458.

Werker, E., & Ahmed, F. (2008). What do nongovernmental organizations do? *Journal of Economic Perspectives, 22*(2), 73–92.

CHAPTER 3

What to expect with volunteers

The life cycle of disasters

Traditionally, the life cycle of disasters spans multiple phases such as preparedness and planning, response, recovery and mitigation or risk reduction (see Fig. 3.1). The life cycle organizes what disaster and volunteer managers do on a daily or seasonal basis and during a disaster event. Disaster volunteers can participate in any of these phases, but most tend to prefer response time activities, and are likely motivated by media coverage alongside a strong desire to help right away. Disaster volunteers can help during all phases though. Imagine how an even safer world might be created if volunteers supported the work of emergency managers across the full life cycle – here are some ideas of what they could do:

- Preparedness/Planning Phase
  - Take classes in first aid, CPR, and shelter management to be ready to help.
  - Distribute information on area hazards and readiness actions at county fairs and public events.
  - Focus on high risk populations:
    - Teach children how to shelter in place for area hazards.
    - Assist seniors and other high risk populations with creating an emergency kit.
    - Organize a local animal rescue effort and pet sheltering plan.
  - Participate in planning initiatives:
    - Lead colleagues through creating a business continuity plan at work, places of worship, schools, and agencies.
    - Join a local emergency planning committee (LEPC) for a potential hazardous materials incident.
    - Support efforts to design, train, and exercise on response plans.
    - Encourage long-term recovery planning to promote resilience.

Disaster Volunteers. DOI: https://doi.org/10.1016/B978-0-12-813846-5.00003-4

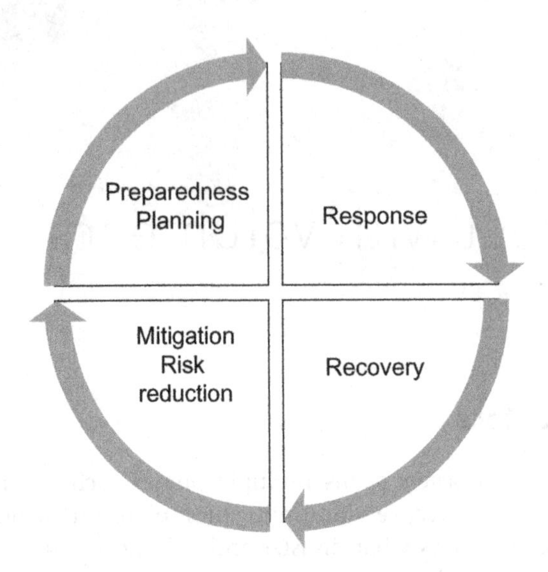

FIGURE 3.1
The life cycle of disasters.

- Mitigation Phase
  - Support mitigation planning to identify risks and help prioritize solutions such as dams, levees, and safe rooms.
  - Preserve local community:
    - Check the homes of highly vulnerable populations for potential risks.
    - Join a local preservation effort to safeguard cultural resources and historic properties.
    - Conduct a mitigation assessment at work or worship locations to identify where disaster risk reduction measures could be implemented.
  - Raise funds for mitigation projects in the homes of people at high risk (e.g., secure bookcases in earthquake prone areas) or for a community-wide mitigation effort (e.g., building elevations in floor prone areas).

Recovery phase

- Focus on repairs and reconstruction:
  - Raise money to fund unmet needs.
  - Organize, distribute, and manage donations with an eye toward items needed for long-term displaced families to eventually return home.

- Tear out insulation, sheetrock, and damaged furniture so that repairs and reconstruction can begin.
- Paint interiors and exteriors of homes, put on roofing materials, and clean and repair furniture and household goods.
- Address deferred maintenance (like weakened roofs, windows, or doors) that put people at risk in future disasters – and fix the issues to prevent future damage.
- Raise money for local groups working to help those in need.
- Support volunteer management including lodging, feeding, and serving those who volunteer during recovery.

- Offer professional expertise and services and:
  - Repair electrical and plumbing problems and supervise repairs and construction efforts (being mindful not to displace local workers).
  - Provide interim medical, optometric, psychological, and dental care, especially for low-income families, seniors, and people with disabilities until local health care facilities re-open.
  - Offer pulpit supply or other interim workers so that affected employees can launch their personal recovery.
- Address under-served needs:
  - Volunteer in child care for displaced families so they can focus on recovery, being mindful to join credentialed and certified efforts.
  - Help small businesses in need of repairs but lacking in funds, especially woman- and minority-owned businesses.
  - Focus on those who fall through the gaps of available funding who would not recover without help – seniors, single parents, historically oppressed populations, refugees, and other groups at high risk for continued displacement.
  - Show up to rebuild homes, playgrounds, and communities and to make them accessible across age groups and abilities.
  - Offer foster care to displaced pets and help build a dog park.
  - Back farmers by joining a hay lift for hungry livestock or working their fields.
- Support trauma healing efforts:
  - Sit with a survivor and just listen to their stories.
  - Secure training for case management and help people through the recovery process including finding resources, navigating aid bureaucracy, and overcoming what feels like big obstacles to going home again.
  - Participate in a range of secular and spiritually-based efforts to support those who are struggling, being conscientious about respecting their beliefs as part of the healing process.
  - Organize outings, fun events, celebrations, house blessings, and more to generate relief and recreation for survivors.

Disasters present an array of opportunities to volunteer and serve humankind. While the response period may feel like the most urgent, in reality, help is also needed during other times. Imagine what might happen if we pushed the collective energies of volunteers into preparedness, planning, mitigation, and recovery? More lives would be saved, less damage would occur, and those harmed by disasters could pick up their lives and livelihoods sooner. Families would benefit, communities would thrive, and economic losses would be reduced.

While the non-response periods are not as visible or perhaps as dramatic or impactful – volunteers can make a difference in such phases, out of the media attention, as a volunteer in the everyday life cycle of disaster management. Such seemingly mundane work, like painting a wall, truly matters to people who just want to go home. The labor provided by disaster volunteers can also be especially helpful to emergency and volunteer managers who may be the only paid employee in their agency. Knowing who will volunteer and how to use them in these phases represents the first critical steps that emergency and volunteer managers need to take to leverage the social capital that volunteers bring to a disaster.

Types of disaster volunteers

Media coverage shows clearly how disasters imperil people, seemingly leaving them without homes, jobs, and resources. Thanks to the wide reach of traditional and social media, viewers assume that they should respond because societies, communities, and families raise people to engage in such pro-social help (Kaniasty & Norris, 1995). We teach our young to behave altruistically, support such behaviors through faith traditions and family modeling, and encourage those around us to participate in civic efforts. Harnessing the subsequent volunteer potential represents a significant task for disaster managers, mission team leaders, and volunteer coordinators to address the wide array of disaster tasks.

To anticipate, receive, and deploy volunteers, managers should anticipate that two kinds of disaster volunteers usually appear: spontaneously unaffiliated or unplanned volunteers (often called SUVs) and volunteers affiliated with formal organizations focused on disasters (Britton, 1991; NVOAD, no date).

Unaffiliated volunteers

The most likely kind of volunteer to arrive is the well-intentioned but unaffiliated SUV who arrives without planning to do so. Such a phenomenon has been known for a long time – and is referred to as "convergence"

(Fritz & Mathewson, 1956). Why do volunteers self-deploy or converge on a disaster scene? One explanation says that people assume that disasters create needs and that people need help right away. In response, both *material* (unsolicited donations) and *personal convergence* (unsolicited volunteers) may occur (Fritz & Mathewson, 1956; Neal, 1993, 1994).

Material convergence, (e.g., canned goods, bottled water, used clothes) surprisingly overwhelms disaster sites – and often goes un-used. Yes, such a phenomenon really does happen. In contrast, the best action to undertake is to send money rather than used clothing or canned goods. Why? Because locals can use financial contributions to re-stimulate the area's economy and to buy specific items that people need such as hearing aids, disability-specific assistive devices, prescription medications, food for various nutritional demands, and more. Material convergence, like personal convergence, can be truly problematic.

Personal convergence means that people (sometimes hundreds or thousands) arrive with little notice. Imagine both the potential positive and negative impacts of such arrivals in an area where floods have taken out bridges, debris has closed roads, and storms have downed electrical lines. Logistically, will unaffiliated volunteers be able to enter the area? Will they be aware of the potential dangers so they can move about safely? SUVs may not have clothing or gear to protect themselves from tornado-shattered glass or hurricane-strewn hazardous waste. They may or may not arrive with sufficient water, food, or plans for shelter – or with gloves, boots, shovels, and other necessary equipment. Local areas, already impacted by the disaster, will now need to deal with the arrival, organization, and deployment of such volunteers or send them away – on top of the disaster. Keeping volunteers may require a range of amenities including bathrooms, communication, coordination, medical support, supervision, food, water, security, and housing.

Conversely, do unaffiliated volunteers provide help? Yes, under the right circumstances when recipient communities anticipate and manage their arrival (see Table 3.1). Being ready for SUV arrival matters, including abilities to provide training and oversight to increase volunteer safety (Sauer, Catlett, Tosatto, & Kirsch, 2014). With such care, those arriving may then commit for longer-term efforts. After the Black Saturday bushfire in Australia, spontaneous volunteers created a longer term effort called BlazeAid to support damaged rural areas (Whittaker, McLennan, & Handmer, 2015). As a form of citizen participation, spontaneous volunteering has significant potential for positive impacts in areas of need, if anticipated and managed well.

In a response time period, when SUVs are most likely to appear, people arrive highly motivated to help, willing to undertake difficult work,

Table 3.1 Pros and cons of spontaneous unaffiliated volunteers (SUVs).

Pros	Cons
Motivated	May create local impacts
High level of initiative	Not as prepared
Willingness	Too many to be used efficiently
Caring and compassion	May go un-used
Availability	Need amenities (bathrooms, food, water)
Therapeutic effect	Typically lack background checks
Financial benefits	May lack needed training for disaster tasks
	May need equipment
	Not as coordinated
	Not as organized
	Displace local workers
	Potential exposure hazardous materials
	Safety concerns/Injuries
	Are strangers in a traumatized area

and dedicated to relieving the burdens of traumatized survivors. The initiative and enthusiasm of SUVs and other volunteers produces a *therapeutic effect*, which occurs when beneficiaries feel the weight of the disaster lift (Barton, 1970; Phillips, 2014, 2015). From listening to people to performing manual work, volunteers produce social and psychological benefits beyond their physical labor for survivors and, quite often, for themselves as well (Steffen & Fothergill, 2009; Thoits & Hewitt, 2001). The therapeutic effect, while noticeable in the response period, may be even more pronounced during long-term recovery when many agencies and volunteers have forgotten the damaged community (Phillips, 2014).

Affiliated volunteers

In contrast to SUVs, affiliated volunteers serve through experienced disaster organizations. Affiliated volunteers may range from local people organized and trained by the International Red Cross/Red Crescent or in Community Emergency Response Teams (CERTS, see Phillips, Mwarumba, & Wagner, 2012; Simpson, 2001) or arrive in timed, deployed units like a Medical Reserve Corps (MRC, see Phillips et al., 2012). One prominent organization is the Red Cross, which operates in many nations like the Canadian Red Cross, the British Red Cross Society, or the Qatar Red Crescent Society. Most Red Cross/Red Crescent units include paid staff and extensive cadres of trained volunteers. Their volunteers stand ready to offer first aid, mass care (e.g., food, water, shelter), case management, and psychological care immediately after a disaster occurs.

As an example, the County Health Administrators and Oklahoma Commissioner of Health activated Medical Reserve Corps units with animal response teams after a devastating tornado damaged the town of Moore in 2013. Called a County Animal Response Team or CART, volunteers joined pre-planned and predictable shifts in a designated site. Such a carefully coordinated effort rescued hundreds of animals, conducted medical exams, provided transportation to shelters or veterinary facilities, photographed lost pets for social media, and searched for owners. Shelter volunteers bathed, fed, and cared for lost pets and livestock, reunited pets with humans, and eventually coordinated adoptions. The careful, pre-planned effort saved and comforted hundreds of traumatized animals and used volunteer labor effectively.

When disasters occur, help will be needed. But, organized, pre-planned, and coordinated volunteer help, associated with experienced disaster organizations, best serves damaged communities. And, while the immediate response time may seem the right time to deploy, the majority of help will be needed with the long-term recovery process of rebuilding, long after the media have moved on and recovery becomes less visible to the public (Table 3.2). Experienced disaster organizations know this, and offer experience in how to assign volunteer teams. Affiliated volunteers represent the best management strategy to leverage the kindness of disaster volunteers, especially those arriving from outside the damaged area. Imagine the challenges of trying to enter a place like Port-au-Prince, Haiti, after the 2010 earthquake. The earthquake damaged the water and airports. Debris clogged the roads. Water problems and food insecurity increased. Physical safety, threatened by aftershocks, worsened. Simultaneously, challenges erupted from saving survivors entombed in collapsed buildings to moving them offshore into floating

Table 3.2 Pros and cons of affiliated volunteers.

Pros	Cons
More likely to train volunteers	Some may seek publicity
Organized deployment	May not match local needs
May arrive self-sufficient	May need financial support
Mission specific	Logistical coordination
Oversight/Supervision	May need local housing
Flexible	
Experienced	
Networked	
Collaborative	
May have paid staff	
May stay for years	

hospitals for life-saving procedures. Entering the damaged area, in Haiti's capital, proved extraordinarily difficult, even for experienced military units. Despite the short 90 minutes flight from Miami to Haiti, arrival through the airport would take days (Benjamin, Bassily-Marcus, Babu, Silver, & Marin, 2011). Many volunteers arrived without basic equipment or a means to secure their own nutrition and hydration needs and even medical personnel arrived without credentials (Jobe, 2010). Volunteers also require protection from diseases in a disaster area. After the Nepal earthquake of 2015, for example, 53% of an Israeli rapid response team experienced gastrointestinal distress (Lachish, Bar, Alalouf, Merin, & Schwartz, 2016). Keeping volunteers healthy and safe means that volunteers serve more effectively and do not unduly burden locals.

Not surprisingly, disaster focused organizations and their affiliated volunteers serve as the steady, predictable backbone at disaster sites worldwide. In the U.S., the National Response Framework includes Emergency Support Function (ESF) 6 where disaster organizations coordinate to deliver mass care and address human needs. The leadership of ESF 6 includes the Voluntary Agency Liaison (VAL), a federal position tasked with facilitating and coordinating the entry and services of disaster organizations. Imagine the benefits of being in an affiliated organization when ESF 6 convenes, including:

- Acquiring resources such as phones, computers, information, meetings, and meeting space.
- Hearing briefings on the disaster as it unfolds.
- Accessing useful services including immunizations, food, water, credentialing, maps, parking, and entrée to disaster sites.
- Participating in needs assessments and coordinating who does what.
- Synchronizing deployment of volunteer teams with adequate supervision and support.
- Connecting specific needs to organizational missions and assets.
- Avoiding duplication of services to promote efficient use of volunteer resources.

Which situation would you prefer to manage? The arrival of an unknown number of volunteers eager to help or the organized deployment of trained and credentialed volunteers targeting specific needs? It is very likely that, as a disaster or volunteer manager, you will become involved with both. Embedded within each type of volunteer lies significant social capital (Nakagawa & Shaw, 2004), capable of being leveraged. Within each, and preferably through affiliated organizations, people with specialized or professional skills can be tapped to help.

Professional volunteers

Remember the example given earlier of psychological care provided by Red Cross or Red Crescent volunteers? Although hugs from volunteers may be needed, psychological care for traumatized disaster survivors should be undertaken by well-prepared and experienced professionals. Similarly, medical and dental professionals should deliver care based on their expertise. Architects, engineers, and construction experts should be involved in reconstruction requiring code compliance and safety checks. Electricians and plumbers represent the best people to install or repair damage to utilities. In short, professional volunteers can provide a considerable amount of well-qualified help, which can alleviate financial burdens on local homeowners, renters, and small business owners. They can also support needed services when disasters disrupt such functioning. For example, the Great East Japan earthquakes of 2011, that caused a tsunami and nuclear power station accident, pulled veterinarians into animal rescue, shelter, and medical care (Tanaka, Kass, Martinez-Lopez, & Hayama, 2017).

Both advantages and disadvantages exist for professional volunteers (see Table 3.3). Their expertise can be leveraged to put quality efforts into place due to their certifications, training, and licensing. Professionals generate a high level of competent work to address acute needs. The example from Haiti earlier meant that the first sets of volunteers in needed experience in extricating people from risky areas, treating acutely traumatized survivors with medical and psychological wounds, and managing massive temporary shelter sites. Countries from around the world as well as established organizations sent teams to do just that. Thus, professional volunteers can play a vital role in delivery of crucial survival services, as Australian nurses did in the aftermath of the 2004 tsunami. Though most lacked language skills and relevant immunizations, nurses willingly responded and served ably (Arbon et al., 2006).

Table 3.3 Pros and cons of professional volunteers.

Pros	Cons
Training, certification or licensing	License, Certifications may not be local
Eager to learn	May not be credentialed for affected area
Eager to contribute	Could supplant local professionals
Can play a key support role	May not be available when needed
Can often address unique needs	May be available for specific times
High level of professional work	Could be place-bound due to their work
Familiar with the work that needs to be done	Students do have to go back to school
May have experience in work needed	

Para-professionals or professionals in training may also become involved where laws and standards permit. Students in the Disaster Management Program at Hesston College (U.S.) learn reconstruction, how to work with local recovery committees, and case management during summer field placements. Or, students may become involved as a brand new effort. Nursing students, for example, provided support after a Japanese earthquake, helping to decrease suicides among elderly survivors (Kako & Ikeda, 2009). Paralegals may be needed to help with paperwork preparation, documents, and more.

Some downsides exist, though, and should be considered when planning for volunteer teams of professionals and especially with professionals-in-training. Affected areas may require certifications specific to their own rules, such as surgeons on medical teams or construction contractors – and approvals could take some time without pre-disaster agreements in place. Further, professionals may not always be available at the time of need, resulting in a delayed arrival. Or, they may be available for a specific time period and require replacement such as when students or professionals have to return home. Another consideration is that outside help could displace locals who need such work. Some communities, such as in New Orleans after hurricane Katrina, required the use of local contractors during reconstruction to prevent the displacement of area professionals.

Given their potential value, how do disaster and volunteer managers learn of and work with such professional volunteers? They may do so in several ways through:

- Pre-established agreements such as nursing associations, medical organizations, or an organization like Doctors Without Borders or Veterinary Medical Assistance Teams.
- Civic organizations and clubs that provide services like the Lions who provide eye glasses and use mobile clinics for people with needs linked to their missions.
- New offers that need to be reviewed for credentialing, liabilities, and usefulness. Architecture firms, for example, have redesigned shelters, homes, and small businesses after disasters. One such organization, Build Abroad, launched efforts to rebuild schools after the 2015 Nepal earthquake.
- Participating in efforts like the Inter-agency Standing Committee built on United Nations General Assembly Resolution 46/182. The resolution led to the United Nations cluster system, where participating agencies shoulder responsibility for service delivery in health, logistics, shelter, sanitation and related areas.

Regardless of one's role as an SUV or an affiliated volunteer, or as a professional, para-professional, or student, people arrive to help survivors. It is valuable to know, then, that survivors will also be part of the volunteer effort.

Survivors as volunteers

If the purpose of disaster volunteerism is to help others, why would survivors need to help? Survivors may actually benefit from volunteering in order to sort through what has happened to them, to regain a sense of control, and to see a way out of tragedy. By volunteering, they may gain perspective on their own losses and feel more connected to others experiencing the same situation. By helping, they may alleviate personal isolation and enjoy contributing to a community that is recovering and healing. Survivors from an earthquake in Marmara, Turkey, experienced post-traumatic growth in part from volunteering for disaster preparedness activities (Karanci & Acarturk, 2005). Doing so can make people feel safer as they face what may linger as a continuing hazard.

Further, all disasters occur in a particular context with local languages, customs, practices, beliefs, and ways of doing things. Locals – including survivors – know their local context, culture, and people. To demonstrate, most Long Term Recovery Committees (LTRC) in the U.S. operate at the local level. Such committees are best chaired by people who come from the affected community because they know people and places. After hurricane Katrina, it wasn't unusual for locally flooded survivors to lead LTRCs. Survivors raised funds, coordinated relief and recovery efforts, and took care of their neighbors. In Mississippi, one local committee chair knew who had not applied for help – and successfully brought in new clients whose homes were repaired or rebuilt by non-local volunteer teams. In New Orleans, a local pastor vouched for an outside agency that the local community did not know or trust – and, again, volunteer teams came in to rebuild that home (Phillips, 2014).

Survivors can help themselves, and often with unexpected outcomes. In Gujarat, India, after the 2001 earthquake, women came together to work on textiles to rebuild their livelihoods. As they worked, the women began to encourage each other to send their daughters to school. In time, some of the women ran for local offices (Lund & Vaux, 2009). Turkish women participated in a similar effort where they created dolls for tourists after an earthquake. They became so successful they opened a doll factory and secured a contract with the Ministry of Tourism, thereby generating a completely new industry (Yonder, Akçar, & Gopalan, 2009).

It is not unusual that survivors also become the first responders, and do so for days or weeks until help arrives. They pull people from debris, tend to injuries, call for help, comfort the traumatized, and assist police, fire, emergency, and volunteer managers. They also lead recovery groups, write grant proposals, and feed and house incoming volunteers while rebuilding their own homes and businesses. It is their home, and their community – and many want to give back. By incorporating survivors into the volunteer effort, it may be possible to help those affected as well as outsiders who need to understand local context and culture, a need particularly true when people volunteer in other countries.

International disaster volunteers

Imagine that a terrible disaster has struck a community about 500 km from where you live. How would you go about helping? How would you get there? It is quite a distance, so where would you sleep or eat? Can you buy gas for your vehicle? Once you arrive, where will you go? Are all of the areas safe yet? Did you bring the right clothes for the work and the climate?

Now, magnify those challenges through the possibility of serving or managing volunteers five thousand miles away in another country (see Table 3.4). Do you have passports and, if needed, entry visas? Does your team have the proper immunizations? Are locals expecting you and your team? How familiar are they with your organization and what kind of learning curve should be anticipated while they get to know and trust you? What about your equipment and other resources, including health care and safety needs? Will you be able to get them out of your country and into the disaster zone? Are you planning to arrive self-contained or do you expect the locally affected area to support your team? Are you prepared to reimburse the locals for your needs? Do you have an evacuation plan in case the emergency occurs again or worsens? International volunteer teams need to be ready to enter gently and appropriately, and to minimize their impact on locals. Just wanting to help is not enough. By responding to

Table 3.4 Pros and cons of international disaster volunteers.

Pros	Cons
Highly motivated	Distance
May have significant resources	Language
Often organization-based	Culture
Willing to travel	Knowledge of local policies
	Potential to displace locals who need work
	Donor-driven approach
	May need local support
	May not coordinate but should
	May arrive unannounced and should not

local needs with local people providing guidance and insights, volunteer efforts can be more effective.

Consider, for example, the challenges in Sri Lanka after the 2004 tsunami. Arriving organizations and volunteers assumed support would be available from locals. However, pre-disaster conflicts in Sri Lanka complicated locating and delivering services. Disconnects with the country's bureaucracy slowed efforts, made worse by internal corruption. Donors and organizations also made assumptions about what would be needed including fishing resources. Further, distribution processes, coupled with internal bureaucratic challenges, increased inequities among fisherpeople and caused concern about overfishing. Women in particular lost ground due to an influx of food and garments traditionally produced by local women (Mulligan & Shaw, 2011). Doing the wrong thing for the right reasons fails stricken communities and generates problems. Resulting friction between outsiders and locals can undermine use of resources, which is the last thing that should happen in a disaster (Benjamin et al., 2011). Collaborating with locals, and understanding the local context, best leverages knowledge, resources, donations, and volunteers (Cuny, 1983; Jasparro & Taylor, 2011; Mulligan & Shaw, 2011; Schreurs, 2011).

Social scientists advise all of us to consider two potential stances when encountering other cultures and ways of life. Ethnocentrism represents one of those stances, defined as judging another culture from the perspective of one's own. Too often, people approaching another culture may find themselves puzzled about another way of life. Rather than embracing or understanding other ways of interacting, eating, sheltering, or rebuilding, an ethnocentric person may condemn, misunderstand or undermine how locals do things. People have reasons for why and how they do things. For example, consider that post-disaster housing efforts can easily fail when outsiders do not heed local preferences. After the tsunami in Sri Lanka, donors offered homes that used gas cooking devices instead of firewood, which people could afford and access. Donors also built kitchens without spaces for fishing gear in an area where people earned their livelihoods from the sea (Karunasena & Rameezdeen, 2014). Sometimes called the donor driven, rather than owner-driven approach, ethnocentrists believe or assume they know better than the survivors. Such a phenomenon is not unique to Sri Lanka. Volunteer efforts in Haiti revealed that it is better to "learn and listen rather than do and tell" to be most effective (Welty & Bolton, 2017, p. 124).

In contrast, volunteers who rely on a culturally relative approach try to understand a process, procedure, interaction, or policy from the perspective of the local culture. In a time of trauma, people need familiar ways of doing things and respect for their way of life. They also need help specific to their context and environment. Showing up with inappropriate donations, reconstruction practices, or interaction styles can further traumatize survivors and slow their recovery. Culturally relative approaches rely on indigenous knowledge, a hallmark of best practices for disaster relief and recovery (Prober, O'Connor, & Walsh, 2011). To illustrate, consider the partnered efforts of *Living Waters for the World* and *Solar Under the Sun*. Their approach? To work with locals to identify appropriate sites and train them on how to install and repair solar-powered water treatment plants. Volunteer teams must undergo cultural immersion training including how to empower and support local decision-makers, which they have done in Haiti since the earthquake. With over thirty systems now established in Haiti, locals have benefited from a mutually respectful partnership, saving countless lives by providing clean, affordable drinking water (see http://solarunderthesun.org/projects/haiti.cfm). By working with people rather than for people, and by working within the cultural context rather than by making assumptions about needs, international volunteer teams can make more of a difference.

Types of disaster organizations

Where can volunteers find affiliated organizations for disaster work? Similarly, where should volunteer and emergency managers look for experienced help? A number of "umbrella" organizations provide links into disaster-focused organizations that train and deploy affiliated organizations. In the U.S., the National Voluntary Organizations Active in Disaster (www.nvoad.org) serves that function. Approximately two-thirds of the NVOAD members come from faith-based organizations (FBOs) across a full range of affiliations. Many such FBOs have been active in disasters for decades, bringing with them experienced volunteers, dedicated training programs, and resources. The United Nations cluster approach, and the Inter-agency Standing Committee (IASC), represents a similar international body where governmental and non-governmental organizations may work to address a range of needs in humanitarian crises.

Another way that emergency and volunteer managers can look for disaster-affiliated or relevant aid is to use a disaster organization typology (Dynes, 1974; see Table 3.5). The typology, which arose out of systematic research in

Table 3.5 Organizations in disasters (Dynes, 1974).

	Old structures	New structures
Old tasks	Established	Expanding
New tasks	Extending	Emergent

disaster settings, identifies four main types of organizations. These include *established* disaster organizations like the emergency management agency. In a disaster, they take on traditional or old disaster tasks within established or existing structures like the fire department. A second type, *expanding* organizations, also address familiar disaster tasks but increase their size and capacity to do so, such as when the Red Cross expands by adding volunteers to a shelter. A third type, the *extending* organization, may move in to take on the new tasks of debris removal – such as when a construction firm loans equipment or extends company resources. Finally, an *emergent* organization newly appears ("emerges") in a disaster. Typically, these emergent organizations identify or address unmet needs in new ways and with new organizational structures to do so. An example might be when a group forms to advocate for people left out of the recovery, like single parents, non-English speakers, or people with disabilities. By looking for this range of four types of organizations, emergency and volunteer managers can build relationships and foster a strong foundation for disaster volunteer management.

Certainly, disaster specific organizations offer the best route into disaster volunteering. Established and expanding organizations like the Red Cross/ Red Crescent, the Salvation Army, or similar non-governmental organizations have more than a century of experience. With in-place training programs and paid staff, they can organize and deploy volunteers efficiently and appropriately. Most faith traditions also have disaster or humanitarian assistance wings (Phillips & Thompson, 2013). The five major faith traditions – Christianity, Islam, Judaism, Buddhism, and Hinduism – have all operated in disaster contexts for a long time. Within each faith tradition, particular denominations funnel volunteers into disaster sites. For example, Presbyterian Disaster Assistance has been training, funding, and supporting volunteer efforts for decades. Their experience, coupled with specific internal missions, means that a well-anticipated arrival and roll-out of volunteers can be used well. Many faith-based disaster organizations concentrate on repairs and reconstruction for the homes of historically socially vulnerable populations (e.g., seniors, people with disabilities, single parents, see Phillips, 2014). UMCOR has carefully developed case management procedures and trains people to shepherd survivors through a recovery process.

Voluntary organizations and associations that people belong to in their communities, outside of disaster interests, can be tapped to serve as extending organizations or can provide a route for individuals interested in volunteering. Civic organizations may be well-prepared to help in general or very specifically. Rotarians, for example, organize around the motto "Service above Self" and frequently help in a variety of ways from fund-raising to physical labor. Other organizations, like Habitat for Humanity, may extend volunteers into specific needs such as construction. Community-based organizations can also offer expertise such as when a club or school that uses sign language helps out in a shelter. Colleges and universities can also extend student assistance from programs that support social services, office administration, computer technology, or emergency management.

The business and corporate sector can be brought in too, with the caveat that such help should be focused on specific needs rather than on generic donations. Businesses, like home construction businesses, might be able to lend employees with some already having such efforts in place. Larger corporations might be able to loan space, donate resources, and raise money for smaller businesses, homes, schools, and playgrounds. In the 2013 tornado mentioned earlier, a home construction store donated space, wireless connectivity, parking, and power to the Oklahoma animal response team. Business sectors link organizations to resources including fund-raising or high-tech capabilities. One of the most favorite services provided after disasters comes from companies that sell laundry detergent, when their mobile washer-dryer units appear near the disaster zone.

New, emergent organizations and groups also appear after disasters occur. The tsunami in 2004 resulted in one, called the Happy Hearts Fund. Started by a tsunami survivor, the effort to date has built 150 schools after disasters in Indonesia, Thailand, Haiti, Mexico, the Philippines, and Peru. As another example, long term recovery committees and similar groups represent emergent groups that form after a disaster, the majority of which address unmet needs (Rodriguez, Trainor, & Quarantelli, 2006). Team Rubicon began as an emergent organization, comprised of military veterans who organized to rapidly deploy emergency response teams. Now an expanding organization, Team Rubicon volunteers assist with debris management, home repair, medical support and – yes – also with mitigation efforts.

Conclusion

In short, a wide array of organizations can be identified to help during a disaster. Wise emergency and volunteer managers identify them well in advance, offer disaster specific training, and establish memoranda of understanding about how to help. Volunteers interested in helping should register with an organization, secure training, and stand ready to join disaster volunteer efforts. A full array of volunteers – affiliated or unaffiliated, lay or professional, domestic or international, expanding or emergent, need to be anticipated, organized, trained, and deployed. The remaining chapters walk readers through how to do just that.

References

Arbon, P., Bobrowski, C., Zeitz, K., Hooper, C., Williams, J., & Thitchener, J. (2006). Australian nurses volunteering for the Sumatra-Andaman earthquake and tsunami of 2004: A review of experience and analysis of data collected by the Tsunami Volunteer Hotline. *Australasian Emergency Nursing Journal, 39*, 1–8.

Barton, A. (1970). *Communities in disaster: A sociological analysis*. Garden City, NY: Anchor Books.

Benjamin, E., Bassily-Marcus, A., Babu, E., Silver, L., & Marin, M. (2011). Principles and practices of disaster relief: Lessons from Haiti. *Mount Sinai Journal of Medicine, 78*, 306–318.

Britton, N. (1991). Permanent disaster volunteers. *Nonprofit and Voluntary Sector Quarterly, 20*(4), 395–415.

Cuny, F. (1983). *Disasters and development*. Dallas, TX: Intertech.

Dynes, R. (1974). *Organized behavior in disaster*. Newark, DE: University of Delaware, Disaster Research Center.

Fritz, C. E., & Mathewson, J. H. (1956). *Convergence behavior: A disaster control problem*. Special Report prepared for the Committee on Disaster Studies, National Academy of Sciences, National Research Council.

Jasparro, C., & Taylor, J. (2011). Transnational geopolitical competition and natural disasters: Lessons from the Indian Ocean tsunami. In P. Karan, & S. Subbiah (Eds.), *The Indian Ocean tsunami: The global response to a natural disaster* (pp. 283–300). Lexington: University Press of Kentucky.

Jobe, K. (2010). Disaster relief in post-earthquake Haiti: Unintended consequences of humanitarian volunteerism. *Travel Medicine and Infectious Disease, 9*, 1–5.

Kako, M., & Ikeda, S. (2009). Volunteer experiences in community housing during the great Hanshin-Awaji earthquake, Japan. *Nursing and Health Sciences, 11*, 357–359.

Kaniasty, K., & Norris, F. (1995). In search of altruistic community: Patterns of social support mobilization following hurricane Hugo. *American Journal of Community Psychology, 23*(4), 447–477.

Karanci, N., & Acarturk. (2005). Post-traumatic growth among Marmara earthquake survivors involved in disaster preparedness as volunteers. *Traumatology, 11*(4), 307–323.

Karunasena, G., & Rameezdeen, R. (2014). Post-disaster housing reconstruction: Comparative study of donor vs. owner-driven approaches. *International Journal of Disaster Resilience in the Built Environment, 1*(2), 173–191.

Lachish, T., Bar, A., Alalouf, H., Merin, O., & Schwartz, E. (2016). Morbidity among the Israeli Defense Force response team during Nepal, post-earthquake mission, 2015. *International Society of Travel Medicine, 24*(2), 1–5.

Lund, F., & Vaux, T. (2009). Work-focused responses to disasters: India's Self Employed Women's Association. In E. Enarson, & P. Chakrabarti (Eds.), *Women, gender and disaster: Global issues and initiatives* (pp. 212–223). New Delhi, India: Sage.

Mulligan, M., & Shaw, J. (2011). Achievements and weaknesses in post-tsunami reconstruction in Sri Lanka. In P. Karan, & S. Subbiah (Eds.), *The Indian Ocean tsunami: The global response to a natural disaster* (pp. 237–260). Lexington: University Press of Kentucky.

Nakagawa, Y., & Shaw, R. (2004). Social capital: A missing link to disaster recovery. *International Journal of Mass Emergencies and Disasters, 22*(1), 5–34.

National Voluntary Organizations Active in Disaster. (No date). *Managing spontaneous volunteers in times of disaster: The synergy of structure and good intentions*. http://PointsofLight.org/Disaster. Accessed January 15, 2008.

Neal, D. (1993). *Flooded with relief: Issues of effective donations distribution. Cross training: Light the torch* (pp. 179–182). Boulder, Colorado: Natural Hazards Center, Proceedings of the 1993 National Floodplain Conference.

Neal, D. (1994). Consequences of excessive donations in disaster: The case of Hurricane Andrew. *Disaster Management, 6*(1), 23–28.

Phillips, B. (2014). *Mennonite Disaster Service: Building a therapeutic community after the Gulf Coast storms*. Lanham, MD: Lexington Books.

Phillips, B. (2015). Therapeutic communities in the context of disaster. In A. Collins, S. Jones, B. Manyena, & J. Jayawickrama (Eds.), *Hazards, risks, and disasters in society* (pp. 353–371). Amsterdam: Elsevier.

Phillips, B., Mwarumba, N., & Wagner, D. (2012). The role of the trained volunteer. In Leonard. Cole, & Nancy. Connell (Eds.), *Local planning for terror and disasters* (pp. 177–188). Wiley and Sons.

Phillips, B., & Thompson, M. (2013). Religion, faith and faith-based organizations. In D. Thomas, B. Phillips, W. Lovekamp, & A. Fothergill (Eds.), *Social vulnerability to disasters* (pp. 341–366). Boca Raton, FL: CRC Press.

Prober, S., O'Connor, M., & Walsh, F. (2011). Australian aboriginal peoples' seasonal knowledge: A potential basis for shared understanding in environmental management. *Ecology and Society, 16*(2), 12. http://www.ecologyandsociety.org/vol16/iss2/art12/ (accessed January 5, 2016).

Rodriguez, H., Trainor, J., & Quarantelli, E. L. (2006). Rising to the challenges of a catastrophe: The emergent and prosocial behavior following Hurricane Katrina. *Annals of the American Academy of Political and Social Science, 604*, 82–101.

Sauer, L., Catlett, C., Tosatto, R., & Kirsch, T. (2014). The utility of and risks associated with the use of spontaneous volunteers in disaster response: A survey. *Disaster Medicine and Public Health Preparedness, 8*(1), 65–70.

Schreurs, M. (2011). Improving governance structures for natural disaster response: Lessons from the Indian Ocean Tsunami. In P. Karan, & S. Subbiah (Eds.), *The Indian Ocean tsunami* (pp. 261–282). Lexington KY: University of Kentucky Press.

Simpson, David M. (2001). Community Emergency Response Training (CERTs): A recent history and review. *Natural Hazards Review, 2/2*, 54–63, May 2001.

Steffen, S., & Fothergill, A. (2009). 9/11 Volunteerism: A pathway to personal healing and community engagement. *The Social Science Journal, 46*, 29–46.

Tanaka, A., Kass, P., Martinez-Lopez, B., & Hayama, S. (2017). Epidemiological evaluation of cat health at a first-response animal shelter in Fukushima, following the Great East Japan earthquakes of 2011. *PLoS One, 12*(3), 1–15.

Thoits, P., & Hewitt, L. (2001). Volunteer work and well-being. *Journal of Health and Social Behavior, 42*(2), 115–131.

Welty, E., & Bolton, M. (2017). The role of short term volunteers in responding to humanitarian crises: Lessons from the 2010 earthquake. In M. Holenweger, M. Jager, & F. Kernic (Eds.), *Leadership in extreme situations* (pp. 115–130). Cham, Switzerland: Springer International Publishing AG.

Whittaker, J., McLennan, B., & Handmer, J. (2015). A review of informal volunteerism in emergencies and disasters: Definition, opportunities and challenges. *International Journal of Disaster Risk Reduction, 13*, 358–368.

Yonder, A., Akçar, S., & Gopalan, P. (2009). Women's participation in disaster relief and recovery. In E. Enarson, & P. G. D. Chakrabarti (Eds.), *Women, Gender and Disaster* (pp. 175–188). New Delhi, India: Sage.

CHAPTER 4

Recruiting and training disaster volunteers

Recruiting volunteers

Although spontaneous unplanned volunteers will probably always appear in all kinds of disasters, wise managers also recruit volunteers to help. Why? One reason is because spontaneous volunteers tend to serve during the immediate response time period after the disaster. In reality, disaster-damaged communities will need the most help during the longer-time period of repairing and rebuilding. Another reason is because disasters are not equal opportunity events. Low income families tend to be harder hit due to the affordability of safe structures and will need help from volunteer organizations (Fordham, Lovekamp, Thomas, & Phillips, 2013; McCoy & Dash, 2013). In addition to low income households, those who may face an uphill battle to return to normal include single parents, senior citizens, some people with disabilities or in poor health, and people whose work schedules make it difficult to find free time (Dash, 2013; Davis, Hansen, Kett, Mincin, & Twigg, 2013; Peek, 2013; Thomas, Newell, & Kreisberg, 2013). Volunteers can help people get back home by providing free labor, raising money, securing rebuilding donations, and dedicating themselves to those most likely to be harmed and least likely to be able to recover.

Thus, when media attention wanes after the immediate response time, and volunteer numbers drop off, pre- and post-disaster recruiting of volunteers will make a difference. This chapter starts with an overview of issues related to recruiting volunteers and should be of interest to emergency managers, volunteer coordinators, government officials, and leaders of civic and non-profit organizations concerned with disaster relief. Volunteers will also find content useful as they prepare to serve. This section starts with understanding who volunteers as a foundation for understanding where to recruit even more help.

Disaster Volunteers. DOI: https://doi.org/10.1016/B978-0-12-813846-5.00004-6

The demographics of volunteering

After hurricane Katrina, I interviewed a survivor in a home rebuilt by volunteers. She described the volunteers: "they were like ants.... they were everywhere, they never got tired, and they just kept coming." With some estimates nearing 600,000 volunteers in just the first year after the hurricane, Katrina set a record and a standard for disasters (Corporation for National & Community Service, 2006). Volunteers continued to come for nearly ten years, usually through organized volunteer teams from experienced disaster organizations. And within the mass of people who went to the Gulf could be found an incredible diversity: men, women, adults, children, senior citizens, people with disabilities, unskilled and skilled helpers, racially and culturally diverse volunteers, faithful from all religious traditions, and people from all educational and income levels and around the world. That diversity contains some of the keys to recruiting volunteers (Meyer et al., 2016).

For example, it makes sense that age influences volunteering. Parents and middle-aged employees deeply involved in raising families and earning livelihoods may find it harder to volunteer, especially if it requires traveling hundreds of miles away for a week or more. Not surprisingly, then, volunteer numbers fall highest among younger and older age groups. Managers and leaders looking to recruit volunteers might want to consider where to access these age groups, such as schools and universities or retirement communities (Devaney et al., 2015). To recruit them might involve appeals specific to their age groups and by using the communication portals they rely on to acquire information like social media, email, or more traditional ways to reach an audience like radio or television. Agencies and organizations could also create programs that speak to age groups, like spring break volunteer trips for college students (including for academic credit) or recreational vehicle sites that lure retirees into serving.

Cohorts within age ranges also seem to matter, with groups like baby boomers (born 1946–1964) more interested in volunteering. Some studies suggest that younger generations may be less interested in volunteering, though a more nuanced approach may be needed to understand their interests. For the millennial generation (born between 1980 and 1995), two contrary research threads suggest that either they are more or less interested in volunteering (e.g., see Ertas, 2016; Ng, Gossett, & Winter, 2016). However, volunteer turnout seems to be higher for causes they care about, especially for those already holding a public service orientation (Ertas, 2016). Recruiters would thus be wise to look for places where millennials already work or volunteer in public service settings and build a compelling case to volunteer after a disaster. This strategy may be particularly effective if linked to workplaces that support such efforts (Ertas, 2016).

Indeed, recruiting needs to be nuanced to be effective. While it is true that baby boomers are more likely to volunteer, additional demographics influence turnout. To start, women tend to volunteer more frequently than men (Ruiter & DeGraaf, 2006). Not surprisingly, younger women volunteer more frequently than middle-aged women but that demographic seems to increase again when women grow older, revealing that age, cohort, and gender interact to influence volunteering. Women appear more likely to volunteer if they had a good experience early in life tied to their belief systems. That was the case among activists who supported the civil rights movement, which yielded a higher turnout as those women became older (McAdam, 1988). Recruiters wanting to attract those most likely to say yes might want to look for older women with a history of early volunteering who express strong beliefs in causes they value.

Another aspect to consider is that traditionally gendered volunteer roles have changed dramatically, thus recruiters should look widely for volunteer capacity. Today, women manage construction crews, wire electrical panels, or operate fire suppression crews (Beatson & McLennan, 2005) while men offer emotional support, manage work site offices, and cook. If gender does become a barrier to recruitment on volunteer sites, one strategy is to inspire women-only rebuilds. That has been successful for Habitat for Humanity which inspired Our Women Build efforts (see https://www.habitat.org/volunteer/near-you/women-build). Thus, attending to the demographic of volunteering matters when trying to recruit people to disaster sites. In addition to gender, the impacts of race and ethnicity should also be considered.

Studies of race, ethnicity, and volunteering have suggested differing rates of volunteering (Musick, Wilson, & Bynum, 2000; Wilson, 2000). At first look, it seems that people are more likely to volunteer within their racial group. Once again, though, demographic complexity has to be considered. For example, income and educational levels seem to neutralize racial differences in turnout. One study found, in comparing older white and African American volunteers, that resources mattered (Gonzales, Shen, Wang, Martinez, & Norstrand, 2016). Higher rates of volunteering for African Americans were positively influenced by financial assets, health, and neighborhood resources. For older white volunteers, age, marital status, financial assets, and health resources were associated with formal volunteering. In reaching out to recruit volunteers, it would be wise to offer a range of volunteer opportunities so that people could opt in as resources provide. In addition, recruiters might want to consider how to increase resources for volunteering so that everyone – no matter their income level or resource base – could help. Faith based settings tend to do this well, by offering a range of possibilities like funding disaster mission teams, serving on the team, or praying for the team (Phillips & Jenkins, 2009).

Inclusion also extends to people with disabilities, who are eager to volunteer but are often overlooked (Bruce, 2006). Sometimes seen as recipients rather than as producers of aid, organizations often fail to invite people with disabilities (Shandra, 2017). Such social exclusion also denies the benefits of volunteering to people with disabilities, who actually volunteer more annual hours and weeks than people who do not face disability challenges (Shandra, 2017). Further, it is clear that people with disabilities provide crucial services in disasters. For hurricane Katrina, the Louisiana School for the Deaf involved students in debris removal, website assistance, and translation services. Following flooding in Houston, Woodhaven Baptist Deaf Church used its facilities as a volunteer site and provided offices and space to handle relief efforts.

Therefore, asking everyone to volunteer makes sense in order to tap into the full range of abilities and energies available after a disaster. Efforts should be made to reduce barriers to participation by making volunteer centers accessible, providing transportation, and offering a variety of volunteer activities (National Council on Disability, 2009; Shandra, 2017). Managers can look for opportunities to involve people with disabilities in all phases of the life cycle of disaster management including the immediate response time. People with disabilities should also be considered for a full range of tasks, not just those specific to their particular circumstances. Strategies also include integrating disability organizations, agencies, and educational institutions into training for disaster volunteering to provide a natural route into disaster volunteering (National Council on Disability, 2009). Disability organizations, for example, should be highly-integrated into a VOAD, COAD, NGO or INGO effort as well as part of Long Term Recovery Groups.

Finally, what seems clear across all demographics is that people volunteer most frequently when someone *asks them* (Ertas, 2016; Ruiter & DeGraaf, 2006; Wilson & Musick, 1997). Asking people directly matters, because it provides an opportunity to tailor the invitation to causes and beliefs that people value and to address the specific and often overlapping demographics that influence turnout. Recruiters could leverage such efforts further by tasking key agents of socialization, particularly peers, families, friends, and faith networks, with inviting people to volunteer. Recruiters can find agents of socialization in places where people congregate out of shared interests like work places, civic clubs, and worship locations. The social connections that people share within such places enhances the potential that people will say yes, so recruiters should leverage those ties.

Social relationships and social networks

One place to find such connections is within faith traditions. Religious faith influences volunteer turnout, because people build meaningful connections that attract them to serve (Lam, 2002). Faith traditions also offer consistent messages about caring for those less fortunate including those struck by calamity. Worship-goers with regular attendance are also more likely to serve, perhaps because they hear and value those messages about service (Nelson & Dynes, 1976; Ruiter & DeGraaf, 2006) Worship-goers are also more likely to say yes if a member of their faith network asks them to serve. Regular attendance likely influences those relationships, building connections and belief systems that people value and find desirable as a way to extend themselves into the broader community.

Faith represents just one way in which social networks embrace specific interests, (Maki & Snyder, 2017). People often organize into communities of interest around shared experiences or interests such as carpentry, quilting, sports, or accounting (Boughton, 1998). By identifying a community of interest, recruiters may be able to attract people to volunteer, such as carpentry guilds for reconstruction, quilters to create warm blankets or public art, accounting firms to help with financial management of a disaster organization, or a sports club that puts together a 5k fundraising race (Enarson, 2000; see also http://disasterquiltingproject.com/, last accessed June 21, 2019). By asking people to pursue their interests in disasters, recruiters may be able to attract skilled volunteers who bring their friends with them to serve. And disasters provide opportunities for many interests: animal rescue, reconstruction, health care support, fundraising, organizing, case management, transportation, translating, donations management, environmental restoration, legal aid, historic preservation, and so much more. The goal of emergency and volunteer managers, then, is to recruit volunteers ready to serve by appealing to the social networks that share common interests.

Recruiting volunteers, then, requires understanding the science of volunteering. Research tells us that paying attention to people's specific demographics matters. By considering how to open doors to people from varying genders, ages, races and ethnicities, interests, and abilities, a broader set of prospective volunteers can be attracted to serve. Both within and across those demographics, recruiters need to discern interest-based social networks and social relationships that can bring in not only individual volunteers but potentially larger groups of like-minded people. Above all, recruiters need to ask people, to invite them to serve, and to provide a good space in which to do so. Once recruiters have volunteers interested, it will be necessary to prepare them to serve – which will also help managers to retain them over time.

Training volunteers

The good news is that many disaster volunteers can be trained quickly and then sent out to be helpful. But many situations also require a higher level of training, such as with bioterror attacks (Clizbe, 2004) or child care (Peek, Sutton, & Gump, 2008). The Church of the Brethren, for example, provides child care workers in disaster situations. Such help is often badly needed, so that parents and caregivers can deal with paperwork, launch cleanup, secure resources like food and clothing, and transition into alternative housing. To work with children, though, requires a high level of training around traumatic events. To insure that properly credentialed volunteers participate in the child care setting, the Church of the Brethren requires 27 hours of training as well as background checks. The training focuses on the physical and emotional needs of children who have experienced trauma. Certified volunteers then go to disaster sites, shelters, or summer camps with a comfort kit to engage children in age-appropriate activities. Since 1980, their efforts have sent 3100 volunteers to serve 87,800 children in 230 disasters (Peek et al., 2008; see also www.brethren.org, last accessed June 26, 2019). The example of in-person training required to prepare disaster child care workers represents one way to train people. In the next section, we look at the full array of training strategies followed by legal and safety considerations. A concluding section then pairs the emphasis on recruiting a diverse set of volunteers with the diversity found among survivors they will serve, and how to prepare for interactions across cultures and contexts.

Training strategies

Many organizations provide training well in advance of a disaster, which is the ideal way to prepare volunteers to serve (more on this in coming chapters). Organizations like the Red Cross/Red Crescent provide both in person and online training to secure basic certifications for general mass care work. Courses can also be taken in first aid and CPR. Higher level work, like shelter management, requires additional levels of training to understand relevant policies and procedures coupled with experience and oversight. Volunteers can also learn more generally from disaster organizations or those that provide insights, like Church World Services. Their free, online webinars promote understanding of the disaster recovery process and who does what (see https://cwsglobal.org/our-work/emergencies/webinars/, last accessed August 20, 2019). Professional-level work, like psychological care, requires disciplinary certifications, examinations, board credentials, and approvals well in advance of sending a volunteer to help. Volunteers should seek out such predisaster training in order to be ready to provide the best help available.

Nonetheless, many people will arrive on the site unannounced and eager to help. While 'just-in-time' training serves as the most common form of preparedness for such spontaneous volunteers, it can vary from in-person training at a volunteer center or on-the-site training for particular activities like cleanup or painting. Many disaster tasks involve repetitive efforts, and that kind of training can be pre-made and provided electronically at a volunteer reception center or online. An example might be debris removal and sorting. In many places, it is essential to sort through debris so that it can be reduced effectively before going to a landfill. Thus, volunteers can be trained with videos to sort for woody debris to be incinerated, plastics to be recycled, or hazardous materials that must be handled and disposed of by personnel trained in using proper protective equipment and safety protocols (Brickner, 1994; Brown, Milke, & Seville, 2011; Hayashi & Katsumi, 1996; Karunasena, Amaratunga, Haigh, & Lill, 2009). Volunteers can view such videos at home, while en route to the disaster, or at a central staging area although hazardous materials handling must be done by certified personnel. Videos can also be helpful when volunteers want to review a particular recommendation for an onsite action.

Another option for just-in-time training is to create a specific training space within a volunteer reception center. For example, health care emergencies use this approach to prepare volunteers to go to separately located points of distributions (Aakko, Weed, Konrad, & Wiesman, 2008). Volunteers might be trained to route traffic through an immunization line, or in how to complete required health care paperwork. They may also be given an overview of how a point of distribution operates including protective care for volunteers as well as the organizational and command structure. People who care for animals displaced by disaster would benefit from training, ranging from how to approach and handle an injured or frightened pet, ways to feed or medicare animals, strategies for cleaning and grooming muddied or oiled creatures, and how to assist with daily care.

A final consideration is that people learn differently. What works for one person might not work for another. A younger volunteer who has grown up with social media and personal cell phones might prefer video tutorials while an older volunteer might like to learn in a group of people with similar interests. People also learn by reading and studying like professionals who must earn certifications to conduct specific post-disaster care, work on utilities, or treat animals. But much routine disaster work comes from direct experience. Some volunteers will learn the skill quickly while others may need repetition and practice with supervision. Knowing the set of volunteers available and how they learn best may facilitate training. Some sites will acquire this information through a quick survey either online or on paper, and then assign people into appropriate teams suited to their experience, interests, and need for supervision. Longer-term work sites often use display boards with projects and work teams. As an example, many disaster organizations welcome their one-week mission teams on Sunday night when they do a survey of volunteer interests and skills. A project leader will then sort the volunteers into work teams by abilities, assign a supervisor, and send them out to take on the tasks at hand. Those tasks may change daily as people learn skills, how to work together, and as new tasks emerge. This kind of training coordination takes place usually within an experienced disaster organization who shoulders responsibility for supervising the volunteers appropriately. This kind of supervision is essential to keep people safe, to make sure that local rules are followed, and to reduce risks for everyone including the possibility of legal action.

Legal considerations

Liability for volunteer efforts remains somewhat unclear throughout many countries. Laws also vary, making it difficult to protect all involved in a volunteer response including individuals and organizations (Eburn, 2003). While lawsuits around volunteerism remain uncommon, and "Good Samaritan" laws exist in a number of places like the U.S. or Australia to protect volunteers (Whittaker, McLennan, & Handmer, 2015), lawsuits still occur. In a study of 24 organizations, three reported that they were sued and one reported that a spontaneous volunteer had sued them (Sauer, Catlet, Tosatto, & Kirsch, 2014).

Organizations and volunteers would therefore be smart to engage in risk reduction actions such as training to reduce problems, buying insurance to protect their organization, staff, and volunteers, and securing legal advice when problems do occur. A number of professionals, particularly health care workers, buy malpractice insurance to cover themselves. It might be wise for professionals like psychologists, veterinarians, physicians, electricians, or plumbers working in a disaster area to do so as a 'just in case' resource. A decision to provide overage may require a careful internal discussion of risk management and financial capabilities. Having insurance will not protect an organization from legal action, but volunteers do remain unlikely to sue or be sued in a disaster context (Saaroni, 2014). Some voluntary organizations do offer insurance coverage for a serious injury that would require medical intervention or emergency transport. In the study of 24 organizations, 18% said that they provided such benefits to cover volunteer injuries or deaths (Sauer et al., 2014). Volunteers may also be asked to provide proof of insurance and emergency travel support before leaving for a disaster site. Such support may be particularly important in remote or international locations should an emergency occur.

Safety first

Risk management is the process of determining whether or not employees, organizations or, in this case volunteers, might be exposed to something that could hurt them (to consider what you might be willing to do, see Box 4.1).

Box 4.1 What are the limits of your volunteering?

Everyone has limits to what they can do or are willing to do as a volunteer. Sometimes, our work and family lives limit what and where we can volunteer. Other times, it is our own personal willingness to take risks. In this boxed feature, we learn about two kinds of volunteering that can risk the lives of a volunteer. One comes from the spread of a potentially deadly pandemic called Ebola that involves health care volunteers. The other addresses the risks associated with confronting terrorism through clandestine activities.

Ebola Ebola – a deadly, contagious disease that requires both professionals and volunteers. Would you be willing to volunteer overseas or even in a hospital in your own country to take care of the ill? Focused mostly in west Africa, the disease spreads fear and has proven difficult to eliminate. The World Health Organization (2019) estimates that in the Democratic Republic of the Congo alone, over 2200 people came down with the disease from August 2018 to June of 2019 and over 1400 died.

Experts recommend careful efforts to prepare professionals and volunteers to help with such outbreaks (Wildes et al., 2014; Center for Disease Control, 2019). First, people knowledgeable and experienced in both humanitarian response and medical protocols should be selected. Those who agree to help should also be provided with ebola-specific information and procedures to follow to reduce exposure and to enhance their abilities to help. Resources should also be provided in sufficient quality and quantity to afford the proper levels of care for patients and protection for health care providers. A plan of evacuation should also be given in a health care volunteer falls ill in order to return them to the highest level of care possible. All who travel to areas of infection should also be screened upon return and followed carefully until any illness outbreak time frames have passed (Wildes et al., 2014; CDC, 2019).

Confronting terrorism In a number of countries, people have put their lives on the line to save historic treasures and cultural relics. During conflicts, people have voluntarily transported ancient manuscripts, sculptures and art, and religious artifacts to safety. In World War II, French citizens moved stained glass windows from Chartres Cathedral into caves to save them from Nazi bombings and thefts. Using farm wagons, they saved the12th and 13th century handcrafted windows which were later re-installed safely. UNESCO later designed Chartres Cathedral as a World Heritage site, representing the best of French gothic art (https://whc.unesco.org/en/list/81/).

With the advent of modern terrorism, numerous heritage sites faced massive destruction. The Taliban destroyed multiple sites deemed offensive to their view of Islam and religion including Buddhist sites and ancient manuscripts by Muslims. Buddhist sandstone sculptures in the Bamiyan Valley in Afghanistan, considered World Heritage sites, have been at particular risk with the Taliban using dynamite to blow them up. In Syria, activists have been working to document atrocities by ISIS and to preserve relics at risk of destruction. Antiquities not destroyed have been sold illegally as a means to fund terrorism acts. Librarians and their family members in Mali spirited 377,000 illuminated and rare manuscripts out of Timbuktu through an elaborate road and water-borne journey. Their efforts saved works from the 1400s and 1500s that revealed African knowledge about astronomy, math, science, and faith. When the French military liberated the city, Al Qaeda militants burned over four thousand manuscripts. Efforts made by the impromptu and largely volunteer effort preserved centuries of early knowledge that would have otherwise been destroyed.

What would you be willing to do in the face of such disasters and threats?

[For more on Syria, see https://www.motherjones.com/politics/2015/03/how-isis-cashes-illegal-antiquities-trade/ (last accessed August 12, 2019). For more on Mali, see Hammer, J. (2016). The bad-ass librarians of Timbuktu. NY: Simon & Shuster.]

Source: Center for Disease Control (CDC). (2019). Ebola recommendations for organizations. https://wwwnc.cdc.gov/travel/page/recs-organizations-sending-workers-ebola, Accessed 21.06.19.

Wildes, R., Kayden, S., Goralnick, E., Niescierenko, M., Aschkenasy, M., Kemen, K., ..., Cranmer, H. (2014). Sign me up: Rules of the road for humanitarian volunteers during the Ebola outbreak. Disaster Medicine and Public Health Preparedness, *9, 1, 88–89.*

World Health Organization (WHO). (2019). Ebola situation reports: Democratic Republic of the Congo. https://www.who.int/ebola/situation-reports/drc-2018/en/, Accessed 21.06.19.

Disasters do cause an array of potential conditions that could injure people from downed power lines to wayward insects and reptiles to debris piles with hazardous materials, glass, or metal that could hurt people. Have volunteers died or become injured during disaster response and recovery efforts? Although most reports of injuries and deaths remain reported by the media, one study found that 43% of the voluntary organizations surveyed (n = 24) had reported injuries to volunteers and that two volunteers had died (Sauer et al., 2014). Thus, safety considerations should be part of any training protocol including both physical and psychological safety.

Physical safety

Physical safety starts with determining if the damaged area is safe to enter, which should be undertaken first by seasoned professionals. In many areas, officials will erect barricades around a damaged area in order to control access which volunteers must respect. Once officials determine that an area can be accessed safely, the type of work that will be done should be discussed. Emergency management officials will probably conduct a damage assessment to identify areas impacted and the kinds of damage sustained, which they usually categorize from minor to destroyed (more on this in Chapter 5). Building officials will probably mark off buildings that should not be entered due to risk of collapse or other dangers. Trainers and managers should respect these protocols to insure safety for volunteers and to alert those who want to help to potential dangers. Volunteers and units sending people to a disaster area should offer safety education to reduce injuries and basic first aid training just in case an accident occurs. Managers might also require that volunteer teams carry basic first aid kits and communication devices in order to call for help as well as information on the availability of local first aid and emergency care sites.

Ideally, communities accepting volunteers will require or provide safety gear appropriate for the work that volunteers take on. Such gear could be as simple as work gloves to sturdy boots or hard hats or as complex as personal protective equipment which comes in several levels up to high hazard exposure. Managers may also want to require volunteers to bring their own gear and to go through a safety check before heading to the volunteer site. For both provided and personally owned equipment, training sites should double-check the gear and make sure that volunteers know how to use it and that such use is required. Even filling and placing sandbags benefits from some technical training. Repairing or rebuilding a home may require various kinds of skills from tearing out damaged walls safely to installing wallboard and carpentry correctly. Training thus requires consideration of a range of potential circumstances in order to make sure that volunteering helps rather than hurts.

Local sites may also be subject to area workplace and safety regulations for which compliance will be mandatory. For example, in New Zealand the Health and Safety at Work Act of 2015 covers some volunteer organizations if they employ people to help with volunteer efforts. In such cases, the organization must make sure that people are safe from injuries, ill health, and possible psychological harm. Volunteer workers must also have safe equipment and safe work systems as well as the training they need to do their work. Volunteer worker sites must have adequate amenities including toilets, water, and places to wash and eat. Organizations also have a responsibility to monitor work sites. Volunteers have a responsibility to follow the procedures outlined and the safety protocols including using protective equipment and reporting any health or safety concerns (see https://worksafe.govt.nz/).

Even with such careful advance preparation and awareness of risks, medical care may still be needed. Volunteers have been killed after coming into contact with downed power lines or from building collapses. Helpers will step on nails or cut their hands on broken glass, get a splinter from woody debris, injure their backs, succumb to the heat and humidity, or become ill completely unrelated to the disaster. Safety concerns also arise from the use of tools. Anyone who has used a hammer realizes the potential to hit a finger by accident. Think next about using a chain saw which can cause even worse injuries (e.g., see Box 4.2). Limbs should certainly be protected carefully, but so should eyesight and hearing due to how wood chips can fly and sound can injure. After Katrina, 6.3% of volunteers reported they had a number of injuries or illnesses from insects to contagious illnesses (Swygard & Stafford, 2009). Or, ponder the problems created by flooding after hurricane Floyd in the U.S. when hog waste from five breached waste lagoons entered waterways not to mention the impacts of deceased animals: 28,000 hogs, 2 million chickens, 750,000 turkeys, and 700 cattle (Schmidt, 2000). Just imagine the kind of protective gear and training that volunteers would need from effluents and carcasses.

Volunteers should certainly practice standard safety and don protective gear only after appropriate training along with careful supervision. In the U.S., the Community Emergency Response Teams require volunteers to take safety gear to every training session and to carry it in a backpack while on site. Some areas may require a higher level such as the various kinds of personal protective equipment (PPE) that may be needed in areas where hazardous materials could be present. Careful training and certification is typically required for most hazardous materials levels which should be used only by certified professionals due to the high level of risk of accidental exposure (e.g., see https://www.epa.gov/emergency-response/personal-protective-equipment).

Safety matters should also be addressed beyond the obvious such as tools. As an example, safety efforts should consider transportation to the work site,

Box 4.2 Chain saw safety

Operating a chain saw is inherently hazardous. Potential injuries can be minimized by using proper personal protective equipment and safe operating procedures.

Before starting a chain saw

- Check controls, chain tension, and all bolts and handles to ensure that they are functioning properly and that they are adjusted according to the manufacturer's instructions.
- Make sure that the chain is always sharp and the lubrication reservoir is full.
- Start the saw on the ground or on another firm support. Drop starting is never allowed.
- Start the saw at least 10 feet from the fueling area, with the chain's brake engaged.

Fueling a chain saw

- Use approved containers for transporting fuel to the saw.
- Dispense fuel at least 10 feet away from any sources of ignition when performing construction activities. No smoking during fueling.
- Use a funnel or a flexible hose when pouring fuel into the saw.
- Never attempt to fuel a running or HOT saw.

Chain saw safety

- Clear away dirt, debris, small tree limbs and rocks from the saw's chain path. Look for nails, spikes or other metal in the tree before cutting.
- Shut off the saw or engage its chain brake when carrying the saw on rough or uneven terrain.
- Keep your hands on the saw's handles, and maintain secure footing while operating the saw.
- Proper personal protective equipment must be worn when operating the saw, which includes hand, foot, leg, eye, face, hearing and head protection.
- Do not wear loose-fitting clothing.
- Be careful that the trunk or tree limbs will not bind against the saw.
- Watch for branches under tension, they may spring out when cut.
- Gasoline-powered chain saws must be equipped with a protective device that minimizes chain saw kickback.
- Be cautious of saw kick-back. To avoid kick-back, do not saw with the tip. If equipped, keep tip guard in place.

Source: Occupational Health and Safety Administration. https://www.osha.gov/Publications/3269-10N-05-english-06-27-2007.html, Accessed 24.06.19.

with properly trained drivers conveying volunteers to avoid vehicle accidents, especially in areas where a disaster has destroyed street signs and traffic lights that keep people safe. Personal safety education should also be provided so that any chance of falling victim to criminal behaviors can be avoided, although the reality is that crime rates fall after a disaster (Brezina & Kaufman, 2008; Fisher, 2008). Perhaps the biggest concern that survivors have is that their belongings will be stolen. This behavior, usually referred to as looting, rarely occurs in a disaster (Auf der Heide, 2004; Barsky, Trainor, & Torres, 2006; Fisher, 2008; Quarantelli, 1994) especially among volunteers. However, while criminal behavior appears to be rare in a disaster, it is wise to be ready for a range of potential legal concerns.

Regardless of the nature or size of the threat, we always want volunteers to come home physically and psychologically healthy (Paton, 1994; Paton, 1996).

Psychological safety and volunteering

Seeing the damage caused by a massive disaster can be disturbing. The sheer magnitude of the event can feel overwhelming, and one's efforts may seem insignificant. It may also feel impossible to know where to start and how to determine who to help. It can be hard to make those decisions as an emergency or volunteer manager, as an experienced disaster organization, or as a volunteer. Leaving the area can be difficult as well, knowing that people remain struggling. Psychological safety, then, includes some risks for depression, anxiety and sometimes more challenging mental health impacts.

What puts people at highest risk for psychological impacts? Studies indicate that exposure and proximity influence how volunteers fare psychologically in disaster work (Thormar et al., 2010). Given that friends, family, and strangers serve as the very first to respond in many cases it may be difficult to avoid being impacted. If someone you know has died or been severely injured, or if you volunteer close to such loss you could become affected as a volunteer. This is called exposure and if people are not adequately prepared to deal with the impacts, they could experience difficulties later. Limited research suggests that younger volunteers may be at higher risk for negative outcomes, probably as a consequence of having limited life experiences in dealing with tragedy (Thormar et al., 2010). What can help is advance training and preparedness for potential exposure. Such training occurs most frequently when people join experienced disaster organizations that prepare their volunteers, especially those that may be involved in the first phases of emergency response. If we want volunteers to go home healthy, and to return again as disaster volunteers, we need to help them process their experiences, and if necessary, to offer coping resources. What is key to recognize is that proximity to exposure may affect a person and to seek help from an appropriate source like religious clergy, employee assistance programs, or a therapist.

Another psychological outcome that arises out of longer exposure has been documented among social workers who report trauma symptoms from working with affected clients. Called "secondary traumatic stress" the outcome results when people listening to those traumatized begin to experience similar symptoms, such as bad dreams, tremors, or anxiety (Baird & Kracen, 2006; Boscarino, Figley, & Adams, 2004). Another kind of reaction might occur, called compassion fatigue or burnout (Berg, Harshbarger, Ahlers-Schmidt, & Lippoldt, 2016). Such outcomes have been experienced by social workers and health care professionals and reasonably could also impact longer-term volunteers. Length of time spent in such situations, degree of exposure to the trauma, having previously unprocessed traumas or minimal coping abilities, a lack of support systems and a lack of training could influence the prevalence and extent of psychological impacts (Norris, Friedman, & Watson, 2002;

Norris, Friedman, Watson, & Byrne, 2002). The lesson from research is to recognize that if someone has had a traumatic encounter, they should reach out to discuss their experience and gain coping skills usable so they may rebound in a healthy manner and continue to volunteer in the future.

The mental health and stress reactions of volunteers should thus be a consideration for managers and volunteers themselves (Dyregrov, Kristoffersen, & Gjestad, 1996). Volunteers should be prepared in advance for what they will see, allowed to process their experiences while volunteering, and then debriefed after they return from a volunteer site. Experienced disaster organizations should provide a range of such support, from pre-developed materials usable on a volunteer site to ideas for how to share the experience upon leaving the field. These materials might range from informal assignments like journaling, art therapy, small discussion groups, or professional debriefing.

Generally, though, the good news is that volunteers secure positive benefits more frequently than negative outcomes. People feel better because they helped, as they realize they have made a difference. Volunteers also confirm that volunteering with people who hold similar values affirms their sense of self and belonging and makes them feel better. So, the good news about psychological safety is that volunteering seems to promote well-being, especially if managers prepare people in advance. Chapter 9 will describe these psychological benefits in more detail. Because volunteers will likely come from a wide array of backgrounds, part of training must involve readying volunteers for the people they will meet and the communities they will serve in order to promote not only help, but aid that results in therapeutic effects for those served (Barton, 1970).

Preparing to serve a diverse community

People come from everywhere to help when disaster strikes, generating a cross-cultural set of interactions that can prove both challenging and enriching. Advance training for the people volunteers will encounter would be advisable. Consider, for example, working with people who have hearing challenges. People who are deaf or hard of hearing represent a very diverse set of people. People who are deaf from birth likely use a specific sign language, belong to strong and vital deaf cultures, and will probably want to be volunteers. In contrast, someone with late-onset deafness may face a different set of challenges including social isolation, lack of access to information, and difficulty understanding what volunteers are saying to them. Different modes of interacting and communicating will be needed, from sign language interpretation within the deaf community to the use of written notes for an elderly person who has lost their hearing. Think about it this way: how would a volunteer ask a homeowner what color of paint they would prefer in their rebuilt home depending on their hearing? Because asking preferences empowers homeowners to feel stronger and more confident, and to move further along in their post-disaster healing process, such communications truly matter. Training can be undertaken to help volunteers work effectively with those they want to serve.

Cultural diversity matters too, because providing aid to people in the context of their own, familiar ways of interacting once again prompts healing and enables survivors to move on (Chester, 2005; De Silva, 2006). The foundation of culture is language, which organizes how we communicate with each other from verbal to non-verbal interaction. The latter may require an understanding of eye contact, handshakes, or appropriate interpersonal distances. Verbal interactions would benefit from knowing the right way to ask a question, with the best intonations, and coupled with the most effective nonverbal way of presenting the question.

Recruiting volunteers who speak the local language may be particularly helpful and should be a requirement internationally. But even learning a few words in a local language can help, because people truly appreciate the effort to communicate across native tongues. A local language teacher or professor or heritage speaker may be able to help train people before they leave to serve. The added value will likely include contextual information around the language including how to say hello coupled with the appropriate way to make eye contact or shake hands. Another element of language comes from the ways that people use local expressions. Such idioms often reflect something important to people's identities. Canadians often add "eh" at the end of a sentence, something that Louisianans heard after volunteers came south across their mutual border. Canadians also learned new expressions and words like "who dat" which came to represent the New Orleans Saints football victory after Katrina but actually originated much earlier in the music scene of the city. As one New Orleans-area homeowner smiled and told me about her cross-cultural interactions with volunteers, "now we just say 'who dat, eh?'" That priceless moment of interaction brought people together across national borders, cultures, and languages to create a new expression that provided a meaningful connection and a lot of smiles.

Culture also influences our ideas of hospitality and accepting help. Homeowners having their homes rebuilt often like to bring food to volunteers including regionally-specific cuisines, which should be accepted. They do so as a means to thank their volunteers, and also as a means to engage in important norms of reciprocity. Sharing a meal matters to people, from a simple breakfast pastry to a full dinner. By creating opportunities to engage in such a shared meal, people can connect across differences and build ties that last a lifetime. Training should always incorporate ways to build such connections.

Conclusion

Emergency and volunteer managers and the disaster organizations that rely on volunteers will need to engage in two critical activities. The first one, recruiting, means that people who need volunteers will have to dedicate efforts to attracting help. Paying attention to the demographics of volunteering matters, because people will be able to volunteer only if it fits within their time and availability. Thus, developing programs specific to age groups might represent a possibility. Recruiters can also influence volunteers to help if they tap into social networks linked to specific interests. Faith communities represent a solid source for recruiters as they provide messages consistent with helping others and also foster social networks of like-minded people. Once recruited, managers and organizations have a responsibility to train volunteers effectively. Such preparedness will increase the physical and psychological safety of volunteers and reduce risks on the volunteer work sites. Training volunteers also has the added benefit of increasing volunteer retention because they feel more ready for what they will face. In the next chapter, we will learn more about managing volunteers particularly after they arrive on a disaster site. Content will enrich this chapter further, particularly just-in-time training.

References

Aakko, E., Weed, N., Konrad, R., & Wiesman, J. (2008). Rethinking volunteer management using a centralized volunteer staging and training area. *Disaster Medicine and Public Health Preparedness, 2*(2), 127–129.

Auf der Heide, E. (2004). Common misconceptions about disasters: Panic, the "disaster syndrome" and looting. In M. O'Leary (Ed.), *The first 72 hours: A community approach to disaster preparedness*. Lincoln, NE: iUniverse Publishing.

Baird, K., & Kracen, A. (2006). Vicarious traumatization and secondary traumatic stress: A research synthesis. *Counseling Psychology Quarterly, 19*(2), 181–188.

Barsky, L., Trainor, J., & Torres, M. (2006). *Disaster realities in the aftermath of hurricane Katrina: Revisiting the looting myth*. Newark, DE: University of Delaware, Disaster Research Center, Miscellaneous Report #53.

Barton, A. (1970). *Communities in disaster: A sociological analysis*. Garden City, NY: Anchor Books.

Beatson, J., & McLennan, J. (2005). Australia's women volunteer fire fighters: A literature review and research agenda. *Australian Journal on Volunteering, 10*(2), 18–27.

Berg, G., Harshbarger, J., Ahlers-Schmidt, C., & Lippoldt, D. (2016). Exposing compassion fatigue and burnout syndrome in a trauma team: A qualitative study. *Journal of Trauma Nursing, 23* (1), 3–10.

Boughton, G. (1998). The community: Central to emergency risk management. *Australian Journal of Emergency Management, 13*, 2–5.

Boscarino, J., Figley, D., & Adams, R. (2004). Compassion fatigue following the September 11 terrorist attacks: A study of secondary trauma among New York City social workers. *International Journal of Emergency Mental Health, 6*(2), 57–66.

Brezina, T., & Kaufman, J. (2008). What really happened in New Orleans? Estimating the threat of violence during the hurricane Katrina disaster. *Justice Quarterly, 25*(4), 701–722.

Brickner, R. (1994). How to manage disaster debris. *C&D Debris Recycling*, 8–13.

Brown, C., Milke, M., & Seville, E. (2011). Disaster waste management: A review article. *Waste Management, 31*, 1085–1098.

Bruce, L. (2006). Count me in: People with a disability keen to volunteer. *Australian Journal on Volunteering, 11*(1), 59–64.

Chester, D. (2005). Theology and disaster studies: The need for dialogue. *Journal of Volcanology and Geothermal Research, 146*, 319–328.

Clizbe, J. (2004). Challenges in managing volunteers during bioterrorism response. *Biosecurity and Bioterrorism: Biodefense Strategy, Practice, and Science, 2*(4), 294–300.

Corporation for National and Community Service. (2006). Katrina by the numbers. Available from https://www.nationalservice.gov/pdf/katrina_volunteers_respond.pdf, Accessed 19.06.19.

Dash, N. (2013). Race and ethnicity. In D. Thomas, B. Phillips, W. Lovekamp, & A. Fothergill (Eds.), *Social vulnerability to disasters* (2nd ed., pp. 113–138). Boca Raton, FL: CRC Press.

Davis, E., Hansen, R., Kett, M., Mincin, J., & Twigg, J. (2013). Disability. In D. Thomas, B. Phillips, W. Lovekamp, & A. Fothergill (Eds.), *Social vulnerability to disasters* (2nd ed., pp. 199–234). Boca Raton, FL: CRC Press.

De Silva, P. (2006). The tsunami and its aftermath in Sri Lanka: Explorations of a Buddhist perspective. *International Review of Psychiatry, 18*(3), 281–297.

Devaney, C., Kearns, N., Fives, A., Canavan, J., Lyons, R., & Eaton, P. (2015). Recruiting and retaining older adult volunteers: Implications for practice. *Journal of Nonprofit & Public Sector Marketing, 27*, 331–350.

Dyregrov, A., Kristoffersen, J., & Gjestad, R. (1996). Voluntary and professional disaster workers: Similarities and differences in reactions. *Journal of Traumatic Stress, 9*(3), 541–555.

Eburn, M. (2003). Protecting volunteers? *Australian Journal of Emergency Management, 18*(4), 7–11.

Enarson, E. (2000). We will make meaning out of this: Women's cultural responses to the Red River Valley flood. *International Journal of Mass Emergencies and Disasters, 8*(3), 39–62.

Ertas, N. (2016). Millenials and volunteering: Sector differences and implications for public service motivation theory. *Public Administration Quarterly, 40*(3), 517–558.

Fisher, H. (2008). *Response to disaster* (3rd ed.). Lanham, MD: University Press of America.

Fordham, M., Lovekamp, W., Thomas, D., & Phillips, B. (2013). Understanding social vulnerability. In D. Thomas, B. Phillips, W. Lovekamp, & A. Fothergill (Eds.), *Social vulnerability to disasters* (2nd ed., pp. 1–32). Boca Raton, FL: CRC Press.

Gonzales, E., Shen, H., Wang, Y., Martinez, L. S., & Norstrand, J. (2016). Race and place: Exploring the intersection of inequity and volunteerism among older black and white adults. *Journal of Gerontological Social Work, 59*(5), 381–400.

Hayashi, H., & Katsumi, T. (1996). Generation and management of disaster waste. *Soils and Foundations*, 349–358.

Karunasena, G., Amaratunga, D., Haigh, R., & Lill, I. (2009). Post disaster waste management strategies in developing countries: Case of Sri Lanka. *International Journal of Strategic Property Management, 13*, 171–190.

Lam, P. (2002). As the flocks gather: How religion affects voluntary association participation. *Journal for the Scientific Study of Religion, 41*(3), 405–422.

Maki, A., & Snyder, M. (2017). Investigating similarities and differences between volunteer behaviors: Development of a volunteer interest typology. *Nonprofit and Voluntary Sector Quarterly, 46*(1), 5–28.

McAdam, D. (1988). *Freedom summer*. NY: Oxford University Press.

McCoy, B., & Dash, N. (2013). Class. In D. Thomas, B. Phillips, W. Lovekamp, & A. Fothergill (Eds.), *Social vulnerability to disasters* (2nd ed., pp. 83–112). Boca Raton, FL: CRC Press.

Meyer, M., Peek, L., Unnithan, N., Coşkun, R., Tobin-Gurley, J., & Hoffer, K. (2016). Planning for diversity: Evaluation of a volunteer disaster response program. *Journal of Cultural Diversity, 23*(3), 106–113.

Musick, M., Wilson, J., & Bynum, W. (2000). Race and formal volunteering. *Social Forces, 78*(4), 1539–1571.

National Council on Disability. (2009). *Effective emergency management: Making improvements for communities and people with disabilities*. Washington DC: National Council on Disability.

Nelson, L., & Dynes, R. (1976). The impact of devotionalism and attendance on ordinary and emergency helping behavior. *Journal for the Scientific Study of Religion, 15*, 47–59.

Ng, Eddy S. W., Gossett, Charles W., & Winter, Richard (2016). Millennials and public service renewal: Introduction on millennials and public service motivation (PSM). *Public Administration Quarterly, 40*(3), 412–428.

Norris, F., Friedman, M., & Watson, P. (2002). 60,000 disaster victims speak: Part II. Summary and implications of the disaster mental health research. *Psychiatry, 65*(3), 240–260.

Norris, F., Friedman, M., Watson, P., Byrne, C., Diaz, E., & Kaniasty, K. (2002). 60,000 disaster victims speak: Part I. An empirical review of the empirical literature, 1981–2001. *Psychiatry, 65*(3), 207–239.

Paton, D. (1994). Disaster relief work: An example of training effectiveness. *Journal of Traumatic Stress, 7*(2), 275–288.

Paton, D. (1996). Training disaster workers: Promoting wellbeing and operational effectiveness. *Disaster Prevention and Management, 5*(5), 11–18.

Peek, L. (2013). Age. In D. Thomas, B. Phillips, W. Lovekamp, & A. Fothergill (Eds.), *Social vulnerability to disasters* (2nd ed., pp. 167–198). Boca Raton, FL: CRC Press.

Peek, L., Sutton, J., & Gump, J. (2008). Caring for children in the aftermath of disaster: The Church of the Brethren Children's Disaster Services Program. *Children, Youth and Environments, 18*(1), 408–421.

Phillips, B., & Jenkins, P. (2009). The roles of faith-based organizations after hurricane Katrina. In K. Kilmer, V. Gil-Rivas, R. Tedeschi, & L. Calhoun (Eds.), *Meeting the needs o children, families, and communities post-disaster* (pp. 215–238). Washington D.C.: American Psychological Association.

Quarantelli, E. L. (1994). *Looting and antisocial behavior in disasters*. Newark, DE: University of Delaware, Disaster Research Center, Preliminary Paper #53.

Ruiter, S., & De Graaf, N. (2006). National context, religiosity and volunteering: Results from 53 countries. *American Sociological Review, 71*(2), 191–210.

Saaroni, L. (2014). Managing spontaneous volunteers in emergencies: A qualitative risk-benefit assessment model for local governments in Victoria, Australia. *Master of Science Thesis*. University of Leicester.

Sauer, L., Catlet, C., Tosatto, R., & Kirsch, T. (2014). The utility of and risks associated with the use of spontaneous volunteers in disaster response: A survey. *Disaster Medicine and Public Health Preparedness, 8*(1), 65–69.

Schmidt, C. (2000). Lessons from the flood: Will Floyd change livestock farming? *Environmental Health Perspectives, 108*(2), A74–A77.

Shandra, C. (2017). Disability and social participation: The case of formal and informal volunteering. *Social Science Research, 68*, 195–213.

Swygard, H., & Stafford, R. (2009). Effects on health of volunteers deployed during a disaster. *American Surgeon, 75*, 747–753.

Thomas, D., Newell, M., & Kreisberg, D. (2013). Health. In D. Thomas, B. Phillips, W. Lovekamp, & A. Fothergill (Eds.), *Social vulnerability to disasters* (2nd ed., pp. 235–264). Boca Raton, FL: CRC Press.

Thormar, S., Gersons, B., Juen, B., Marschang, A., Djakababa, M., & Olff, M. (2010). The mental health impact of volunteering in a disaster setting. *The Journal of Nervous and Mental Disease, 198*(8), 529–538.

Whittaker, J., McLennan, B., & Handmer, J. (2015). A review of informal volunteerism in emergencies and disasters. *International Journal of Disaster Risk Reduction, 13*, 358–368.

Wilson, J. (2000). Volunteering. *Annual Review of Sociology, 26*, 215–240.

Wilson, J., & Musick, M. (1997). Who cares? Toward an integrated theory of volunteer work. *American Sociological Review, 62*(5), 694–713.

Further reading

Drabczyk, A., & Schaumleffel, N. (2006). Emergency management capacity building: Park and recreation professionals as volunteer managers in cross-systems collaboration. *Journal of Park and Recreation Administration, 24*(4), 22–39.

Rogstadius, J., Teixeira, C., Karapanos, E., Kostakos, V. (2013). An introduction for system developers to volunteer roles in crisis response and recovery. In *Proceedings of the 10th international ISCRAM conference*. Baden-Baden, Germany.

Starbird, K. (2012). What 'crowdsourcing' obscures: Exposing the dynamics of connected crowd work during disaster. In *Proceedings, CI2012*.

Studer, S. (2016). Volunteer management: Responding to the uniqueness of volunteers. *Nonprofit and Voluntary Sector Quarterly, 45*(4), 688–714.

Waldman, S., Yumagulova, L., Mackwani, Z., Benson, C., & Stone, J. (2017). Canadian citizens volunteering in disasters: From emergence to networked governance. *Journal of Contingencies and Crisis Management, 26*, 394–402.

CHAPTER 5

Managing volunteers at disasters

"We Need Help"

Disasters bring out the best in many people, who want to help fellow human beings harmed by something they did not cause. As a consequence, managers can anticipate that volunteers will come to help, particularly during the immediate response time period. They do so because media coverage provides images of compelling needs people feel they can meet. Even if they cannot come, people will volunteer from where they are to donate blood, organize donation drives, and raise money. They will go online to crowd-source help and try to volunteer from a distance. The challenge, then, is for managers to anticipate the arrival of volunteers, in person or from a distance, and be ready to use them wisely. For disaster-prone communities that think "we (may) need help", this chapter lays out key actions that a volunteer management plan should address.

Think, for example, about what a disaster can do and how volunteers could help. An EF5 tornado, for example, can destroy massive sections of a community. Hurricane-related storm surge or cyclones can inundate entire neighborhoods and villages. Earthquakes will tumble homes, schools, and businesses. For most disasters, response volunteers may be able to pick up debris, pull out flood-damaged sheetrock, or stack bricks for later use but the real needs will come when residents need their homes back. Repairing or completely rebuilding a home can take days to years. Haitians remained in tent cities for years after the 2010 earthquake, well after most volunteers had left the earthquake-stricken area of Port-au-Prince. Puerto Rican residents faced debilitating recoveries while waiting for funds to rebuild and volunteer teams to become available after hurricane Maria. People still needed help years later. With hazardous materials events, it may be impossible to enter the area safely for years or decades. Such was the case near 2011 in Japan after the multi-disaster, cascading event that involved an earthquake, tsunami, and nuclear plant accident. The same occurred in Times Beach, Missouri (U.S.) when dioxin contamination rendered the area uninhabitable in 1983.

Disaster Volunteers. DOI: https://doi.org/10.1016/B978-0-12-813846-5.00005-8

Both managers and volunteers should also think beyond the structure that needs repairs or rebuilding to the people who lived in those sites. It can be very hard to recover, especially for people who are elderly or single parents. Residents may also have medical needs or challenging disabilities. Families with low incomes may lack the means to rebuild. Renters, especially those in public housing or affordable units, may be displaced for some time as they await repairs by a landlord. While people think that governments help, the reality is that most families affected by disaster rely on insurance, private savings, some government assistance (if they qualify), and volunteer help. Recovery is usually not possible without coordinated assistance from volunteers. Fortunately, a strong backbone of experienced disaster organizations and non-governmental organizations may be available to assist if emergency and volunteer managers integrate them into their disaster planning efforts.

Something else to consider is that most volunteers will want to help during the emergency response time. Yet, the life span of disasters encompasses more than the response period and includes mitigation, preparedness, planning, and recovery. With advance planning, managers can involve volunteers in efforts that mitigate or reduce disaster impacts (like public education efforts), enhance preparedness (like putting together an evacuation kit for people and pets), support response (debris removal from individual homes), and speed up recovery (repairs and reconstruction). Though ideally volunteers should be used for mitigation and preparedness efforts in order to reduce needs during response and recovery, the reality is that most volunteers will find such routine work less exciting. The media rarely cover these phases, and the public generally lacks knowledge of how important it is to help friends, family, neighbors, and workplaces prepare. So, volunteers will most frequently appear during response operations. Savvy managers should try to lure them back when they will need the most volunteers for helping people and places during the longer-term period of recovery, especially those most vulnerable to disaster and least likely to recover without help.

Thus, help is almost always needed to deal with disasters, no matter where a community is in the disaster life cycle. Two kinds of volunteers will arrive, spontaneous unplanned volunteers (see Chapter 6) and those affiliated with experienced disaster organizations (Chapter 7). Both will require various degrees of understanding and support and both will present benefits and challenges. The goal of the manager is to harness the energy and resources coming their way through volunteer activity and be ready to leverage it for the long haul. Accordingly, a process needs to be in place to retain and attract volunteers for long term recovery work. The relationship with those volunteers, often established through experienced disaster organizations, needs to be strong.

The volunteer "Family"

The NVOAD movement described in Chapter 1 represents a successful model for coordination and collaboration of volunteers (see www.nvoad.org, last accessed June 6, 2019, see also Hearn, 2004). In addition to the national level of organization, many cities and states in the U.S. have their own Voluntary Organizations Active in Disaster (VOAD) units. They meet, talk, train, plan, prepare, and deploy as a well-informed, organized, and focused set of organizations experienced in disaster with a ready cadre of volunteers able to work both response and recovery. They integrate with local officials and expedite a more effective volunteer effort.

In 1999, I watched VOAD partners arrive in Oklahoma City, Oklahoma (U.S.) after a major tornado outbreak across the state. The Oklahoma VOAD unit led discussions in concert with the state office of emergency management and incorporated volunteers arriving from various disaster organizations. At the first meetings, an emergency management official identified specific areas and needs that had to be taken on. Organizational representatives, sitting in a room of over fifty people, raised their hands to take on those needs. One of them, the Seventh Day Adventists (a faith-based organization) offered to manage the donations warehouse. As always happens in disasters, truckloads full of unsolicited donations were arriving without a place or means to store, sort, and distribute items. The Adventists, led by a capable and experienced volunteer, set up operations in an empty food warehouse that featured loading docks, massive space, office rooms, and utilities. The emergency management agency diverted large, incoming trucks to that facility where volunteers offloaded and counted the contents while local housing office staff created a database inventory. At the recovery center, a Red Cross volunteer amassed disaster sites where affiliated organizations needed items. Organizations then retrieved needed items from the voluminous warehouse, which had filled to capacity within days. Unused items, mostly used clothing, went out to local Goodwill stores for distribution first to disaster survivors and then to the general public. Goodwill subsequently baled and stored unused clothes in massive bundles for over a year. It took a lot of effort for this aspect of the disaster alone, with almost all of it accomplished through volunteer labor.

Meanwhile, other organizations stepped up to establish specific building warehouses, such as a construction-oriented site near one badly damaged town. Other organizations could then go to the building warehouse for lumber, nails, roofing materials, or related items. Specific organizations also offered their help with key tasks. A religious organization brought in mobile showers. Another faith-based unit offered roofing teams. Others brought in cooking trailers and provided meals to both survivors and volunteers.

The recovery proceeded well across multiple tornadic paths spanning hundreds of miles because of the experience that Oklahoma VOAD and NVOAD partners brought in.

But what impressed me the most – and what I have observed repeatedly in disaster after disaster, was the notion of a "family" of volunteers as introduced in Chapter 1. People who came from out of state knew some of the volunteers within the state, because they had worked together on other disasters. I witnessed joyful moments between disaster-experienced friends, as they willingly faced a daunting disaster scene. The scene at the first meetings looked and felt like people who were at a family reunion. They knew and trusted each other's experience, talent, and gifts and also knew they would be there for the long haul into the recovery period.

I saw the same thing happen after hurricane Katrina in 2004 (U.S.). This time, an experienced Federal Emergency Management Agency Voluntary Agency Liaison (VAL) connected both in-state experienced and inexperienced organizations with out-of-state organizations. The first meeting addressed how to handle unsolicited donations. This time, the Adventists helped out with aid from a non-disaster National Guard unit that sorted items, an Islamic trucking company that moved items from donation sites to the warehouse, and a distant company that set up an electronic inventory system so that organizations could search remotely for needed items. Although the system emerged out of partnerships between experienced and new organizations, it worked well to manage massive donations for an affected area the size of the United Kingdom.

What may be most useful about the NVOAD framework, besides their carefully crafted points of consensus on disaster work, is the specialized expertise within the organizations. Over time, some have become quite skilled in specific tasks like debris removal, case management, health care, or rebuilding. Many devote themselves to people with low incomes. One organization, HOPE, offers specially trained humans and canines to support crisis response and recovery work. Depending on the area impacted and the local needs generated by the disaster, a wide array of organizational resources can be used. Put together, the collaboration among partner organizations can be leveraged to yield a significant, coordinated impact.

To illustrate, NVOAD-affiliated organizations often participate in creating or supporting Long Term Recovery Groups/Committees (LTRGs). Locally-led (often by local volunteers), the LTRG identifies area needs and creates a place where cases can be presented for consideration by these outside, experienced disaster organizations. Imagine that an elderly woman has lost her home due to a massive flood. Caseworkers present her situation at a regular LTRG meeting, where an outside organization offers aid with framing a home elevated

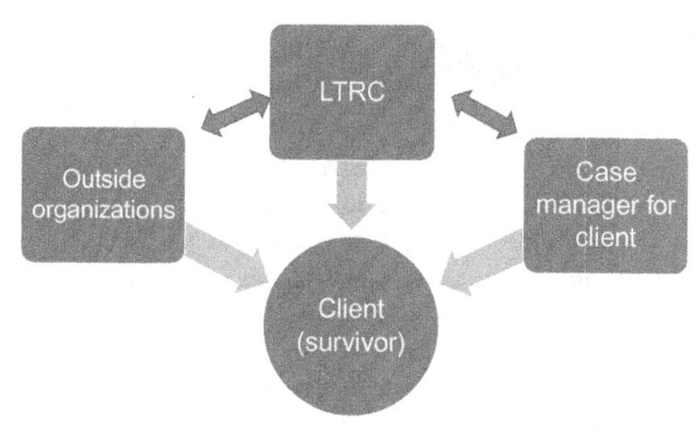

FIGURE 5.1
Discerning survivor needs.

to stay out of future flooding. Another organization then provides roofing materials and crews to do the work. Yet another brings in a mission team to do the interior work on the home, followed by another that paints the unit. Or, in some cases, one organization may offer to do all of that work for a single LTRG client. The most dedicated organizations will set up in a community and work on many homes, bringing in supervised weekly and long-term volunteer teams to do the work. *This* is how people get back home when managers use their volunteers effectively (see Fig. 5.1).

What happened in Japan after the 1995 Kobe earthquake (see also Chapter 1) provides insights into how such disaster volunteer models may need to adapt to local context and culture. Though Japanese have always volunteered, the ultimate responsibility for disaster relief has historically been shouldered by government at various levels. The Kobe earthquake unraveled that approach due in part to disaster's magnitude and impact. Coupled with extensive media coverage, an unprecedented number of spontaneous volunteers turned out exceeding 1.4 million people (Atsumi & Goltz, 2014). Those volunteers then created an emergent group to manage the convergence and ultimately created a national level organization. The volunteers continued to serve in subsequent disasters, learning along the way what worked for them and their people. The volunteer turnout also inspired government action to appoint a national committee of disaster volunteers and new laws to promote disaster volunteerism. In 1995, the original core group of volunteers visited the NVOAD in the United States. Although they now call themselves the Nippon Volunteer Network Active in Disaster (NVNAD), the Japanese organized differently with a focus on emotional recovery. Why not a focus on rebuilding like in the U.S.? It came down to core cultural values of empathy and community (Atsumi & Goltz, 2014). By offering traditional footbaths (which elicits informal talk) and encouraging involvement in local events, survivors became therapeutically reconnected to culture, values, and

Box 5.1 A global relief disaster model

The United Nations has tried to organize humanitarian relief through a cluster model (see https://www.humanitarianresponse.info/en/about-clusters/what-is, Accessed 3.06.19). The core idea is that clusters of humanitarian organizations will coordinate across specific areas such as health, shelter, and early recovery. Each cluster has a lead agency that ideally coordinates at global and country levels, addressing gaps in needs when a disaster occurs (Steets et al., 2010). The approach has been in place since 1991, was revised in 2005, and continues to evolve. Adopted by dozens of countries, the cluster model has worked well in a number of disasters providing leadership and expertise particularly because of paid coordinator positions. However, at the local level, challenges seem to occur where funding, operational coordination, governmental integration, and implementation remain uneven. This appears to be particularly the case in developing nations where local capacity remains low.

community. January 17, the anniversary of the Kobe earthquake, became an annual Disaster Prevention and Volunteer Day (Atsumi & Goltz, 2014). Managing volunteer efforts like these matters, because of the profound impacts that can happen not only for the present-day survivors but for those that volunteers and disaster organizations will serve in the future (see also Box 5.1).

Managing volunteer efforts

To provide a general overview of aspects that should be covered in a volunteer management plan, the remainder of this chapter reviews the process of managing volunteers from their initial arrival through exiting the site, and from the early response period into the long-term recovery.

Making entry

How should people enter a damaged community? Imagine those first hours and days after a disaster: people are heading into your community to help your family and friends, neighborhoods, and businesses. Though help is often appreciated, outsiders also arrive with their own ideas of what to do and how to do it. They come in with their own assumptions, including both well-intentioned and often effective efforts as well as those that go awry. After thirty-plus years of studying such efforts unfold, I know that the influx of volunteers can make a difference when the community is ready for them - and when volunteers prepare well to enter a devastated community (see Box 5.2).

Locally, being ready for volunteers to arrive can be challenging due to the disaster impacts. Regardless, and as Chapter 6 will reveal, spontaneous unplanned volunteers (SUVs) will converge on a disaster site. Because

Box 5.2 Checklist for volunteer team leaders

Are you thinking about going? First, check to see if and when you and your team should deploy. Work with local managers to make sure that your efforts are needed, that your skill sets will be usable, and that you will not unduly impact the local area. A checklist for team leaders should include at a minimum:

- A link into established, affiliated organizations who know you are coming to work with them. This is the ideal way to arrive when invited by local managers.
- A timeline from departure through return.
- A communication list within the team and with organizations at the site and at home, along with a full set of emergency numbers including where to go if an injury or accident occurs.
- Identification.
- An inventory of the skill set on the team from untrained to specific skill sets.
- Tools, protective equipment, and clothing appropriate for the site and climate.
- Training for the anticipated jobs, how a disaster will look (which can be stunning), and safety training.
- Training on how to listen respectfully to survivors, information on their culture(s), and advice for how to connect meaningfully with people who have been harmed by disaster.
- Funding and a plan to arrive self-contained in order to not impact those already struggling with response and recovery.
- Travel logistics including places to buy gas, eat, hydrate, sleep, and shower.
- A plan for how to arrive low-impact and stay that way. Do not expect locals to host your team.
- An intentional mind-set to serve others within the bounds of what you are asked to do.

volunteers will be entering a potentially dangerous area, emergency and volunteer managers will need to think about how volunteers will move safely into the area and return home without injuries. Outsiders should be equally thoughtful as they head toward an area that has been terribly harmed and is still in the early stages of deciding what to do.

Let's start a day or so after the disaster, for example, when an unfamiliar person pulls into a local strip mall parking lot. With a logo on their car door featuring an unfamiliar organization, locals wonder: who is this person? Where did they come from? Then, as these outsiders promise to help, the local resident or manager cannot help but wonder that if they talk the talk, will they walk the walk? Will they work with us or head out on their own? Will they be willing to put up with difficult decisions made locally? Has this organization just shown up for a photo op? In the parking lot example just described, a long-established organization happened to be scouting out locations for long-term housing to bring in volunteer teams over several years. The person wondering "who are these Mennonites?" happened to be a local community leader. As they conversed in the parking lot, the Mennonite explained who they were and what they did, offered a pamphlet, and then accepted an offer to come to a local recovery meeting. Years later, Mennonite Disaster Service was still there working to repair and rebuild homes. What made it work? Both the NVOAD partner and the local recovery leaders took

Table 5.1 Perspectives – what should I be thinking about?.

Emergency managers	Voluntary organizations	Volunteers
When should volunteers arrive?	How many should I send and when?	When should I go?
What kind of work do we have to do? What kind of training do the volunteers need?	What does my organization do well? How have I prepared my team?	What is my skill set? Where can I help best?
What do we need to do to sustain and support the volunteers? Where will they stay, eat, get health care?	How can we arrive self-contained so we do not impact locals?	How can I help without needing help myself?
Who can help us manage volunteers while they are here?	Who should we work with to help with a coordinated and effective effort?	How do I connect with an affiliated organization?

the time to get to know each other, to build trust, and to demonstrate commitment over time on locally-identified needs and people (Phillips, 2014).

From a volunteer perspective, making entry starts with listening to messages about whether to come and, if so, where to start (see Table 5.1). The best advice is always to wait until you are invited so that you can do exactly what the community needs. Give them time to be ready for you. From the manager's perspective, several options exist to organize incoming volunteers, from emergent efforts to well-established volunteer centers. Regardless of the form volunteer management takes, efforts should emanate from a written volunteer management plan crafted by emergency managers, volunteer coordinators, and experienced disaster organizations. After an EF4 tore apart Arkadelphia, Arkansas in 1998, a local resident organized incoming volunteers through telephone, email, and mail. Multiple teams then arrived to help with repairs and rebuilding, in an effort largely coordinated by one person. While this kind of emergent effort still appears (and often works) in many disasters, other ways to manage volunteers have developed. For example, the Volunteer Houston website (http://volunteerhouston.org, last accessed May 14, 2019), features two icons that simply say: "Want to Volunteer?" and "Need Volunteers?" which reveal an architecture designed to link servants with projects. The effort is in place during disasters, such as the 2019 flooding, as well as normal times. As an established framework, it provides a structure through which to match skills with needs. Some sites establish entry points and designate Volunteer Reception Centers, like organizers did after the EF5 tornado in Joplin, Missouri. Other sites may be much more difficult, such as volunteers experienced in trying to get to Nepal after the 2015 earthquake. With roads blocked, making entry proved extremely difficult especially in more remote areas. Drone technology helped with some reconnaissance for voluntary and relief organizations trying to determine where to set up and how to get in. Clearly, being ready to set up a Volunteer Reception Center or entry point serves as the first step that managers and coordinators should plan out including the ability to be mobile with electronic or more tangible resources, like go kits.

Go kits

Existing volunteer centers can be used as the structure for a Volunteer Reception Center or managers can set one up fairly quickly. A good idea is to have a go kit to do so (NVOAD, 2003). Such kits should include:

- *Volunteer Reception Site Information/Maps*. By routing volunteers to a pre-established center or location, efforts can be more carefully coordinated and leveraged.
- *Intake Forms*. Volunteer reception go kits should include a means to accumulate information about volunteer backgrounds, personal and emergency contacts, and skills. Such information can be used to determine how volunteers can be used well and safely. Can be done through an electronic portal, with a site check-in procedure on a computer, or with pre-made forms.
- *Safety Information*. Volunteers should be informed about local conditions and given a training overview on safety matters. Volunteers mean well but they may not understand how disasters create even more hazards. See Chapter 5 for more details.
- *Work Site Information*. Volunteers need to know where they can work. Managers can organize work site information from a simple poster board with handwritten notes to a GPS-based, electronic matching system. For SuperStorm Sandy that damaged multiple states in the U.S., the Long Island (New York) Volunteer Center listed opportunities, skills, and organizations for the Long Island Flood Recovery Effort. Volunteers could sign up online and list their skills in a statewide database for locations needing help.
- *Resource Lists and Locations*. Most affiliated organizations either arrive with their tools or rely on regionally-located warehouses containing chain saws, cleaning kits, protective equipment, and rebuilding tools. Some communities have pre-located tool sheds for various hazards such as snow removal or community gardens that could be re-purposed for the disaster. Home supply stores may be able to rent or donate the use of tools and often have employees with skills sets suitable to supervise volunteer teams. Volunteer management plans should have pre-disaster agreements in place to use them.
- *Information*. Let volunteers know where they can secure additional training in disaster volunteerism such as online courses with the Red

Cross/Red Crescent or Church World Services or through training videos posted online for a specific location (such as how to fill and place sandbags or don protective equipment). Information should also offer a list of open stores, restaurants, and places to sleep as well as the important locations of public toilets and areas to avoid. Tips about the local culture can be helpful including local languages or phrases as well as what people love about their location to help volunteers see beyond the debris to the core of the real community.

- *Thank you notes*. Volunteer Reception Centers should have a way to thank the people who showed up. Volunteers can also be tapped to help write these notes or they can be assigned to school children in the area as a means to help them participate and to process their own experience. Electronic notes can also suffice and be done quickly.
- *Debriefing/Surveys*. A means to gather insights and to assess the volunteer effort is essential, such as a paper survey or an online link to a survey. Those lessons learned should be reincorporated into a volunteer management plan to make the next time easier. Volunteer managers or mission team leaders should also incorporate a debriefing procedure into their efforts, which can range from sharing and processing of volunteer experiences to journaling or more formal efforts provided by professionals.

Messaging with potential volunteers

Any manager, volunteer, or volunteer leader should spend time watching what happens when a disaster unfolds. Because natural hazards happen fairly frequently, it should be easy to spot an event and follow it on social media. For example, just searching Twitter for a hashtag tied to the event should produce a wealth of information. Pay particular attention to the messages given out by managers and by those who want to help. Learn from what you see, as a training ground for how to communicate when disasters happen where you live or want to volunteer.

Increasingly, observers will see that emergency and volunteer managers communicate "do not self-deploy until called" messages in order to organize converging volunteers effectively. NVOAD advocates for such messaging, and encourages volunteers to wait until it is safe to help in an appropriate way (https://www.nvoad.org/how-to-help-volunteer-do-not-self-deploy/, last accessed June 4, 2019). In 2018 storm-related flooding, South Carolina (U.S.) officials asked people to not self-deploy to rescue people in their own boats. Instead, state officials encouraged people to help their neighbors and to work through appropriate organizations (Montgomery, 2018). Such messages should always be communicated through multiple means in order to reach a diverse audience using both traditional and social media.

Advance preparation for communicating includes getting to know traditional media personnel in television, radio, and paper. Relationships matter when disaster strikes, including those with the press. Managers will want to get their messages out quickly and accurately with the help of media. By educating the media on what will be needed and why they want to do it this way, managers can avoid inaccurate messaging that will add to confusion and generate inappropriate responses by the public. Contact local reporters well before disaster and explain the plan to them. Provide them with sample messages. Alert them when a hazard seems imminent and partner with them to put out timely and effective messages.

Social media can be both useful and problematic when disasters occur. Because of the rapid dissemination potential of social media, information (accurate or not) can be delivered to a wide audience. In response to perceived need, volunteers may converge rather quickly and in large numbers, following a call to action from an impromptu media message set up by a well-meaning individual (Bird, Ling, & Haynes, 2012; Starbird, 2013). What emergency managers and voluntary organizations can do to manage such challenges and leverage the potential is to:

- Establish social media tools well in advance of a disaster with a pre-written set of messages to deliver when disaster strikes.
- Integrate social media into traditional communication tools as part of a multi-dimensional public information campaign.
- Provide immediate messages regarding deployment of volunteers including messages to avoid self-deployment.
- Link agreed-upon messages across social media to promote consistency.
- Monitor social media during an event to identify and communicate with people seeking opportunities to volunteer.
- Work with a rumor control center to correct or counter impressions, self-deployment, and wrong perceptions.
- Correct erroneous messages quickly. Encourage people to follow official guidance.
- Offer pro-active messages that direct volunteers into appropriate ways to serve.
- Build cyber-relationships with people and organizations operating across a variety of social media.

Emergency management agencies and voluntary organizations have begun to appoint social media officers or to integrate social media volunteerism into organizational communications. As one example, the North Texas region of the American Red Cross established a Digital Operations Center in 2014

(American Red Cross, 2014). Managers can help to recruit, organize, and deploy volunteers by monitoring social media in an effort to identify possible arrivals of volunteers. By interfacing with these social media sites, volunteers can be advised on deployment time tables, volunteer reception sites, local conditions, resource needs, and safety concerns and can correct impressions about need, safety, and other matters.

Activating a volunteer reception center

When it is time to activate online volunteer registration, a media messaging effort, or Volunteer Reception Center, know that a disaster-stricken area can quickly become a site of massive convergence. Manage such arrivals by:

- Encouraging people to arrive only when you are ready for them.
- Making sure that the site is safe and accessible. Schools, universities, and faith-based sites often serve as such locations as do convention centers or other large, open spaces that afford managers to set up an array of tasks.
- Advertising the volunteer arrival and deployment center via all available media.
- Using an online system to identify needs and assign volunteer teams, which could allow you to avoid opening a Volunteer Reception Center.
- Setting up a clear intake and registration area.
- Creating a badging site or online system that provides identification.
- Providing a space for people to sit while waiting as well as locations for personal hygiene, food and water, information, and training.
- Posting maps on the affected areas or using software to create them. Ideally, maps should identify the site, access points, safety concerns, and specific, confirmed needs (e.g., debris sorting, furniture removal, moving assistance for survivors, etc.). People do like visual information as they move through a registration area, which helps them to begin processing the site they will go to and its needs. Photocopies of maps could be provided or links to an online mapping system. Whiteboards can display current projects as can websites.

Intake

A volunteer intake site will need to be established. Ideally, the site will be located in a Volunteer Reception Center outside of the damaged area in order to provide adequate parking, credentialing, assignments, training, safety education, and transportation into the damaged areas. Locating the site outside the damaged area may also offer the benefit of onsite utilities and restrooms, such as in a convention center, community center, school, or worship site (Aakko, Weed, Konrad, & Wiesman, 2008). To establish intake, managers will need to consider these elements for their volunteer management plan:

- Messaging to direct people to the correct site without bypassing the process that has been established.
- Parking and traffic control including disability access and space for larger vans and buses.
- Signage directing volunteers where to check in.
- A system (electronic or paper) to capture information about volunteers and their skill sets.
- Hard copies of maps and/or apps that offer the same information about the Volunteer Reception Center and the damage areas.
- Multiple intake volunteers to quickly process busloads of arriving spontaneous volunteers.
- A consideration of expedited entry for volunteers arriving with experienced disaster organizations.
- Badging or identification protocols implemented with accuracy and speed.
- A quick orientation video or presentation explaining how the volunteer effort will unfold.
- Volunteers who can assess skill sets quickly and assign people into work teams as indicated by a damage and needs assessment.

Damage and needs assessment

The site should incorporate and deploy volunteers based on information from emergency managers. Trained personnel typically conduct damage assessments in order to determine the numbers of structures and areas that are damaged or destroyed where volunteer effort may be needed (Fleming, 2014). Such information should also include area hazards such as downed utility lines, precarious structures, and hazardous materials to keep volunteers safe. Damage assessments can be done in a variety of ways from windshield surveys to engineering assessments to reviews conducted by drone pilots. Each kind of technique has its pros and cons, with some assessments more accurate than others. Windshield surveys can usually find the areas hit but may not tell the assessor what happened to the integrity of the building structure. That kind of work is best done by structural engineers but that can take time. The damage assessments are usually compiled through an emergency management agency and can result in written lists or detailed mapping. Agencies that have staff skilled in GIS can create maps of the damaged areas. Digital volunteers may be able to do that for a community as well using open source tools.

Ideally, damage assessment should be coupled with needs assessment that specifies who needs what kind of help, where, and when. In a needs assessment, information accumulates over what people need in their homes and businesses, from tearing out sheetrock in a flooded site to helping people move into temporary housing. Needs assessments can also identify vulnerable people or areas where elderly people reside or perhaps a pet refuge that requires assistance. To gather these data, managers could involve an area agency to survey damaged areas in person, via phone call systems, or by using online registries.

By overlaying the damage and needs assessments, a visual pathway can be discerned of who needs what kinds of help, for what purpose, and with what degree of access in the disaster zone. Volunteers or staff with computer skills can be involved in that effort using open source software or systems like GIS. This kind of work can be done from a distance although some level of ground-truthing is advisable for local context.

Training

Pre-departure training makes for a smoother experience upon arrival, and for more effective work. People interested in volunteering or communities that know they will face disasters, could set up periodic opportunities to learn relevant skills. Local community colleges and universities may have construction programs that can lend educators or trainers, as can construction companies, supply stores, and established disaster organizations. Local efforts to build low-income housing, like with Habitat for Humanity, may also afford a chance to develop one's pre-disaster skill set. Volunteers will then arrive knowing what to do well and safely. These kinds of training sites may also represent promising recruiting opportunities to find like-minded people interested in serving people in need.

Not surprisingly, most training takes place on disaster sites. Training specifically at a volunteer site should encompass a range of needed elements including knowledge specific to the type of service to be undertaken, skills that volunteers will use on site, safety protocol, and socialization for interactions with recipients of service. Some volunteers will need specific skills prior to their arrival including background checks, for example, for those who work with children (Peek, Sutton, & Gump, 2008) or professional degrees and certifications to address mental health needs. However, much immediate response efforts include just-in-time training immediately prior to deployment or at the Volunteer Reception Center.

The Volunteer Reception Center, if it is large enough, can hold multiple dedicated sites for training purposes. Parking lots and green spaces can also be used. Ideally, the center will establish sites that link to the damage and needs assessments so that – once past intake – volunteers can be directed to training areas to prepare them for the work they will face. They can learn sandbagging, debris removal, clean-up, and animal rescue procedures within a reasonably quick time before heading to the worksite. Managing volunteers at this point also requires attention to the protective gear they might need which, in an ideal set up, could be checked out once volunteers complete their training along with first aid kits, hydration and nutritional resources, and information about the area including directions (remember, the signs may all be down, though GPS may still work if phones can be charged). All training efforts should start with and insure that people will not be harmed by their service.

Safety

Managing disaster volunteers on a site requires acute levels of care paid to safety concerns. Cleaning up a tornado-stricken house may seem routine but within that house may lie damaged containers leaking pesticides. Disasters also tend to be seasonal, such as cyclones and hurricanes that occur during weather that can dehydrate a volunteer quickly. Affected areas may contain hazardous materials, nails and splintered wood, heavy objects, high heat and humidity, and downed power lines, as well as insects, snakes, fire ants, or other unpleasant conditions (Clizbe, 2004; Neal, 1993, 1994; Paton, 1994, 1996). Volunteer Reception Centers and voluntary organizations should walk volunteers through how to prevent exposure or injury as well as what to do if such an event occurs and where to seek help.

Resources should be provided from insect repellant to sunscreen to, where needed, personal protective equipment. It may be advisable to have volunteers secure immunizations depending on the area in which they travel, provide proof of insurance, pre-arrange for emergency transport, and secure a medical release (Clizbe, 2004). Local public health agencies may need to be ready to offer tetanus or flu shots and to monitor for disease outbreaks that could be brought in by volunteers or compromise their health. Volunteer Reception Centers should let people know where to go for first aid or trauma care which can make the difference between minor and major injuries or even life and death. Safety efforts should also involve professionals who examine sites or personnel for risk. Recall, for example, studies have revealed that asthma, respiratory irritation, and other health risks from exposure at the site of the World Trade Center attack in 2001 (Landrigan et al., 2004; Malievskaya et al., 2002; Wisnivesky et al., 2011). We need to keep people safe. For more ideas on safety, please refer back to Chapter 4.

Transportation

Volunteers need to travel in and around a disaster zone. It would be wise to provide group transportation into and out of a site, especially during the response time period. Volunteer teams should rent vans or similar multi-passenger vehicles with people qualified or credentialed to serve as drivers. Local managers may be able to get such transportation through school buses, public transportation, rental companies, airport vans, and faith-based settings that use vans and buses. Group transportation also alleviates parking and traffic congestion and may reduce damage to the number of vehicles that experience flat tires from navigating debris. As the community transitions from response into recovery and roads clear, voluntary organizations and mission teams should begin providing their own transportation with possible support from local managers.

Drivers should hold the proper credentials to transport groups of people safely. In addition, managers may want to confirm that a particular company or agency loaning out vehicles hold insurance including liability. In areas where known hazards occur repetitively, a formal memoranda of understanding may be negotiated so that such transportation efforts proceed smoothly when a disaster occurs.

Local logistics

Now that volunteers and voluntary organizations have arrived, what happens next? What kinds of local amenities will be needed and how will they be handled? Typically, the first need will be public restrooms. A number of locations could potentially be used from service stations and park restrooms to faith-based settings and portable toilets. A cost will be involved, though, from supplying hygiene supplies like toilet paper, a water source, and soap, to the costs of cleaning up after thousands of volunteers. Some of this cost can be defrayed if volunteers arrive self-contained with their own supplies and if volunteers participate in cleaning restroom facilities they use. Regardless, local managers will need to designate locations and direct volunteers to such necessary sites.

Should volunteers work long hours into the night, they may want a place to sleep and shower. A range of locations can be considered from school and faith-based facilities to public gymnasiums, fitness centers, recreational facilities, or college campuses. Pre-disaster planning should have identified such sites and established memoranda of agreements with the locations including the cost of such housing and the time it can be used. Once again, volunteers should arrive self-contained with sleeping bags or rolls and cots or mattresses. They might also bring tents or recreational vehicles. As volunteer efforts transition out of response and into recovery, longer-term solutions may be sought. In past disasters, volunteers have stayed in tent cities, converted warehouses, community centers, or RV sites. Tent cities are likely to be the least ideal due to the climate and any continuing risk associated with area hazards and the least likely to offer useful amenities like restrooms and showers.

Experienced disaster organizations will rely on their prior experience to identify lodging sites. It is not unusual for a local community to donate a site for temporary use and even consider paying the utilities. Quite often, faith-based settings open their doors to mission teams coming from hundreds of miles away. They may offer not only a place to stay but also shower sites, restrooms, cooking facilities, and laundry options. Experienced organizations may also move into abandoned hotels, churches, and other buildings by renovating them through volunteer work and donor money. In some locations, volunteer housing has been set up for multiple organizations to use on a reservation basis. Businesses that rent space to people with recreational vehicles may also offer discounts or free spaces for longer-term volunteers. Communities and businesses may wish to sponsor such sites by providing amenities at reduced or no charge and sending food to the volunteers.

Volunteer teams and disaster organizations will need other resources, from cleaning supplies to food to paint which may be available through donations or as local purchases. However, given that most donations focus on unneeded items like used clothing and canned goods, it may be necessary to secure rebuilding resources. That work may involve pursuing grants or in-kind supplies from foundations, companies, or individual donors through fund-raising efforts. If the disaster prompted people to donate funds, it would be wise to conserve a significant amount for rebuilding resources that voluntary organizations can leverage through their free labor. Local managers could also create tool banks and warehouses for shingles, paint, lumber, nails, and similar resources. Long Term Recovery Groups will likely need to tap into such locations when volunteer crews adopt their clients. Local recovery committees and leaders will likely need to write grant proposals, conduct fund-raising efforts, and make their case known for additional supplies. After hurricane Katrina, a local leader in Pass Christian, Mississippi (U.S.) traveled to civic clubs around the world. He secured a significant commitment from the Naperville, Illinois Rotary Club to rebuild dozens of homes in his community. In Louisiana (U.S.), the Southern Mutual Help Association helped secure money for the post-Katrina Long Term Recovery Groups. The World Bank supported post-disaster construction efforts after the Nepal earthquake. Organizations like UNICEF, Save the Children, and OXFAM have responded with aid to places like Mozambique, which was severely impacted by flooding in 2019. Resources will be needed and volunteer management plans will need to include efforts to secure those resources in their plans.

Again, arriving self-contained works well but so does spending funds locally. Doing so gets the economy going again and helps to build relationships between outsiders and locals. Shopping local sometimes results in benefits to the disaster organizations working to stretch rebuilding funds. I recall one such moment while researching a voluntary organization. We had stopped at a local hardware store to pick up new paintbrushes. The clerk called into the back of the store: "how much is the two inch paint brush?" The local owner said "$4. Who is it for?" When the clerk replied that it was for an outside, faith-based volunteer group, the manager called out: "Today it's $2."

Volunteers will also need to eat. Experienced organizations will bring their own feeding facilities or rent them locally. A few organizations specialize in feeding volunteers and bring in large canteens and mobile feeding units such as Mercy Chefs. It is not unusual for local businesses to offer discounts at restaurants and grocery stores or to provide entire meals for volunteer teams. Buying local food at grocery stores and restaurants also helps to re-stimulate the local economy and put survivors back to work. However, volunteer teams should recognize that they may need to travel longer distances to secure groceries given the extent of the disaster. That distance will prove particularly challenging in remote sites and where pre-disaster conditions have created food deserts and conditions of deprivation. Feeding people well keeps them fueled for the work they do and represents a wonderful way for disaster organizations and managers to thank their volunteers. It is perhaps not surprising then, that the most important volunteer on a disaster site is the cook (Phillips, 2014).

Interactions among volunteers and locals

Part of restoring hope requires that outsiders recognize, honor, and respect local traditions. People's customs arise out of their culture, defined as their way or life or design for living. While it may perplex outsiders that people live in a rural community subject to floods or in a high-rise building that shakes during earthquakes, it makes sense to the people who live and work there. They like the trees and fields of corn which provide a familiar landscape in which their identity, work, and kin are embedded (Hummon, 1990). Or, the city, with its trendy restaurants and high-brow entertainment venues, provides a stimulating place to work and enjoy life. Cornhole games in the U.S. may be completely unfamiliar but it's as exciting to American mid-westerners as curling is to Canadians. The local fishing hole, the city park with its swans, or the bayou that features dolphins spark a sense of local connection as well.

Managers should invite outsiders to enjoy locals' sense of place and identity, especially as the recovery period lengthens. Consider involving volunteers on sites that reconnect people to their sense of place to aid their recovery physically, socially, and psychologically (Silver & Grek-Martin, 2015). Tell people about the history of the area, and invite them to historic sites and buildings even if they are damaged (for examples, ask locals or visit the National Register of Historic Places in the U.S. or the UNESCO World Heritage Center websites). Tell outsiders about local culture, which is always at the heart of why people love where they live. The languages that locals speak, with its various intonations and dialects, reflects people's heritage and ways to convey meaning. Volunteers can be invited to learn the local languages, whether it is a language different from their own or simply unique words that make sense locally.

Culture also includes food that locals eat. Familiar food, like bagels, pinto gallo, or boiled shrimp provides comfort and connection. Survivors long for what is culturally familiar, for the local jazz music to return or the tamale restaurant to re-open, and for normal life to come back. Invite volunteers to celebrate with locals at re-openings for historic sites, recreational areas, and downtown businesses. Involve outsiders to join in local activities like ball games, ice cream socials, or 5 k runs. Such physical activity helps sore muscles from volunteer labor and distracts locals from the dirty mud that covers their flooded community. People smile when they are having fun, and doing so may be one of the first steps toward normalcy.

Consider creating places where people can interact, such as weekly gatherings between volunteer teams and survivors. Telling their stories can also reduce survivor stress, during interactions in which volunteers play a vital role by listening with an open heart and mind (Pennebaker, 1997; Phillips, 2014, 2015; Steffen & Fothergill, 2009). Such "sympathetic identification" generates therapeutic effects for not only survivors but for volunteers as well (Barton, 1970; Phillips, 2014).

Debriefing volunteers

Studies suggest that volunteers may be affected by what they see, sense, hear, smell or feel on a disaster site (Clukey, 2010). Yet, few organizations provide advance training for what volunteers may encounter at a disaster. Volunteers report that going to disaster sites can be life-changing when they see the extent of damage or listen to other's stories (Pennebaker, 1997). Few people have seen the phenomenal impacts that a disaster can have. Even fewer are likely prepared to handle the emotional impacts on the survivors – and potentially on themselves.

Even well-trained professionals can suffer as well (Dyregrov, Kristoffersen, & Gjestad, 1996). Professionals who fail to process their experiences or who use negative coping skills (such as alcohol or drugs) appear to be at higher risk (North et al., 2001). Thus, it is a good idea to offer some way to process the disaster experience. Debriefing can range from talking about the day's events in a semi-structured group setting to journaling about one's experience. Some organizations bring in a therapist or clergy member while on site to place the events into a framework for thinking about and understanding what volunteers have witnessed. Mission teams or work groups may also want to organize a debriefing session upon their return home in order to place their experiences into perspective. Some volunteers organize presentations to give to their home communities as a way to think through what they have seen, heard, smelled, and experienced. More formal means may include referring people to professional debriefing which walks attendees through discussing and addressing traumatic exposures (Armstrong, O'Callahan, & Marmar, 1991; Everly, 1995; Jenkins, 1998). However, it is important to know that most disaster victims and the majority of volunteers do not develop severe trauma reactions such as Post-Traumatic Stress Disorder (PTSD, see Norris, Friedman, & Watson, 2002; Norris, Friedman, & Watson, Byrne et al., 2002). But volunteering on a disaster site, for many, is life-changing (Mitchell, Griffin, Stewart, & Loba, 2004). Though usually a positive and affirming experience, being ready to help affected volunteers is always wise. Recognizing their contributions also provides benefits through positively affirming their time and energy.

Rewarding and Recognizing volunteers

While volunteering reaps benefits in general as noted in Chapter 1, it is still worthwhile to reward those who serve. Such recognition publicly affirms the act of service and serves as a means to retain volunteers and inspire future efforts. Disaster volunteerism might include these recognitions:

- Group photos and certificates of participation, which can be made available or distributed electronically, perhaps as a download available upon checkout at the Volunteer Reception Center.
- A thank you note, preferably handwritten. Local communities might ask school children to make pictures or write notes to the volunteers or to their organization. Stickers, easily printed off from a computer graphic, can also suffice.
- Local volunteers personally thanking volunteers when they check out at the Volunteer Reception Center. A handshake and a hug go a long way after a day well spent in service to others.
- T-shirts with the organization's logo and disaster event or an embroidered patch or pin specific to the event.
- Special evenings out for a cultural encounter such as local music, poetry, singing or worship.

- Delivery of food to volunteers from a local eatery like pizza, tacos, or gumbo.
- A billboard thanking volunteers, placed on the route where volunteers would exit the affected area.
- Shout-outs on traditional and social media to recognize volunteers in both word and image.
- Awards including plaques or formal recognition at public events.

Conclusion

Ideally, managing disaster volunteers involves all participating parties from emergency and volunteer managers to the volunteers and their voluntary organizations. Throughout the process of deciding to volunteer, and how to manage volunteers, all parties involved should practice mindful, intentional planning. From the initial decision of "should we go or wait?" to the determination to participate in long-term recovery efforts, careful planning and coordination should take place. Volunteers will converge on a disaster in multiple forms, from spontaneous unplanned volunteers to those affiliated with and trained by experienced disaster organizations. While managers usually prefer the latter, it will be necessary to prepare for both. Local managers should anticipate arrivals in advance and design a volunteer reception center that manages arrivals through training, deployment, debriefing, and recognizing efforts. Outsiders should prepare themselves similarly, arriving only when asked and doing so in a manner that is self-contained, equipped for local conditions and cultures, and integrated into a planned system.

References

Aakko, E., Weed, N., Konrad, R., & Wiesman, J. (2008). Rethinking volunteer management using a centralized volunteer staging and training area. *Disaster Medicine and Public Health Preparedness*, 2, 127–129.

American Red Cross. (2014). The American Red Cross: Transforming for the future. Available at https://www.redcross.org/content/dam/redcross/atg/PDFs/Leadership/15202.4-Year_McGovern_Report_PRINT2.pdf, last Accessed 29.10.19.

Armstrong, K., O'Callahan, W., & Marmar, C. (1991). Debriefing Red Cross disaster personnel: The multiple stressor debriefing model. *Journal of Traumatic Stress*, 4(4), 581–593.

Atsumi, T., & Goltz, J. (2014). Fifteen years of disaster volunteers in Japan. *International Journal of Mass Emergencies*.

Barton, A. (1970). *Communities in disaster*. NY: Doubleday.

Bird, D., Ling, M., & Haynes, K. (2012). Flooding facebook – the use of social media during the Queensland and Victoria floods. *American Journal of Emergency Management*, 27(1), 27–39.

Clizbe, J. (2004). Challenges in managing volunteers during bioterrorism response. *Biosecurity and Bioterrorism: biodefense strategy, practice, and science*, 2, 294–300.

Clukey, L. (2010). Transformative experiences for hurricanes Katrina and Rita disaster volunteers. *Disasters, 34*(3), 644–656.

Dyregrov, A., Kristoffersen, J., & Gjestad, R. (1996). Voluntary and professional workers: Similarities and differences in reactions. *Journal of Traumatic Stress, 9*(3), 541–555.

Everly, G. (1995). The role of the Critical Incident Stress Debriefing (CISD) process in disaster counseling. *Mental Health Counseling, 17*(3), 278–290.

Fleming, M. (2014). *An exploratory study of preliminary damage assessment in Northern Illinois.* Master's thesis, Oklahoma State University, Stillwater, OK.

Hearn, P. D. (2004). *Hurricane Camille: Monster storm of the Gulf Coast.* Jackson, MS: University of Mississippi Press.

Hummon, D. (1990). *Commonplaces: Community ideology and identify in American culture.* NY: SUNY Press.

Jenkins, S. (1998). Emergency medical workers' mass shooting incident stress and psychological recovery. *International Journal of Mass Emergencies and Disasters, 16*(2), 181–197.

Landrigan, P., et al. (2004). Health and environmental consequences of the world trade center disaster. *Environmental Health Perspectives, 112*(6), 731–739.

Malievskaya, E., et al. (2002). Assessing the health of immigrant workers near ground zero: Preliminary results of the World Trade Center day laborer medical monitoring project. *American Journal of Industrial Medicine, 42*(6), 548–549.

Mitchell, T., Griffin, K., Stewart, S., & Loba, P. (2004). We will never forget: The Swissair flight 111 disaster and its impact on volunteers and communities. *Journal of Health Psychology, 9* (245), 245–262.

Montgomery, M. (2018). 'Do not self-deploy:' Message from state as boaters help with rescues. https://wpde.com/news/local/do-not-self-deploy-message-from-state-as-boaters-help-with-rescues, Accessed 04.06.19.

Neal, D. (1993). *Flooded with relief: Issues of effective donations distribution.* Cross training: light the torch. *Proceedings of the 1993 national floodplain conference* (pp. 179–182). Boulder, CO: Natural Hazards Center.

Neal, D. (1994). Consequences of excessive donations in disaster: The case of Hurricane Andrew. *Disaster Management, 6*(1), 23–28.

Norris, F., Friedman, M., & Watson, P. (2002). 60,000 disaster victims speak: Part II. Summary and implications of the disaster mental health research. *Psychiatry, 65*(3), 240–260.

Norris, F., Friedman, M., Watson, P., Byrne, C., Diaz, E., & Kaniasty, K. (2002). 60,000 disaster victims speak: Part I. An empirical review of the empirical literature, 1981–2001. *Psychiatry, 65*(3), 207–239.

North, C. (2001). The course of post-traumatic stress disorder after the Oklahoma City bombing. *Military Medicine, 166*(Suppl. 2), 51–52.

NVOAD. (2003). *Points of consensus: Volunteer management.* Washington D.C: NVOAD.

Paton, D. (1994). Disaster relief work: An example of training effectiveness. *Journal of Traumatic Stress, 7*(2), 275–288.

Paton, D. (1996). Training disaster workers: Promoting wellbeing and operational effectiveness. *Disaster Prevention and Management, 5*(5), 11–18.

Peek, L., Sutton, J., & Gump, J. (2008). Caring for children in the aftermath of disaster: The Church of the Brethren Children's Disaster Services Program. *Children, Youth, & Environment, 18*(1), 408–421.

Pennebaker, J. (1997). *Opening up.* NY: Guilford Press.

Phillips, B. (2014). *Mennonite disaster service: Building a therapeutic community after the Gulf Coast storms*. Lanham, MD: Lexington Books.

Phillips, B. (2015). Therapeutic communities in the context of disaster. In A. Collins, S. Jones, B. Manyena, & J. Jayawickrama (Eds.), *Hazards, Risks, And Disasters In Society* (pp. 353–371). Amsterdam: Elsevier.

Steets, J., Grunewald, F., Binder, A., de Geoffroy, V., Kauffmann, D., Kruger, S., ... Sokpoh, B. (2010). *Cluster approach evaluation 2 synthesis report*. Global Public Policy Institute.

Silver, A., & Grek-Martin, J. (2015). Now we understand what community really means: Reconceptualizing the role of sense of place in the disaster recovery process. *Journal of Environmental Psychology, 42*, 32–41.

Starbird, K. (2013). Digital volunteerism during disaster: Crowdsourcing information processing. CHI 2011, Vancouver, BC Canada.

Steffen, S., & Fothergill, A. (2009). 9/11 volunteerism: A pathway to personal healing and community engagement. *The Social Science Journal, 46*, 29–46.

Wisnivesky, J. P., Teitelbaum, S., Andrew, T., BOffetta, P., Crane, M., et al. (2011). Persistence of multiple illnesses in World Trade Center rescue and recovery workers: A cohort study. *The Lancet, 378*, 888–897.

CHAPTER 6

Managing spontaneous unaffiliated volunteers

Introduction

Do not self-deploy is the number one rule of disaster volunteering. Why? Because of all the reasons mentioned so far in this volume and more. When disaster strikes, people want to help people those who have been harmed. And, volunteers come – sometimes by the busloads, often without notice, and wanting to help. However, doing so can present numerous challenges for managers and volunteers alike, from making ingress to the affected area to having the right fit between needs, resources, and help. Communities may not be ready for spontaneous volunteers to appear, often lacking essentials from food, water, housing, parking, transportation, and power to having enough toilets sufficient for a massive influx of volunteers.

We do need volunteers. Studies note that the person most likely to rescue you is a spontaneous volunteer, usually family member or neighbor, because they were nearby when the event occurred (Macintyre, Barbera, & Smith, 2006; Noji, Armenian, & Oganessian, 1993; Sauer, Catlett, Tosatto, & Kirsch, 2014). We have seen this behavior repeatedly occur from small car crashes or house fires to massive tornado outbreaks and active shooting attacks. People willingly take care of each other and render aid when bad things happen. In disasters, survivors report they truly see the best of humanity (Phillips, 2014).

Still, such spontaneous volunteer work is not easy to do. A tornado that damaged rural Alabama in 2019 caused the deaths of twenty-three people in one community. Those spontaneous volunteers faced the most difficult experience of their lives as they searched for the living, helped the injured, and found friends and family who had perished. Chain saw volunteer teams then moved in, amid a debris field generated by a 70 mile long EF4 path. Volunteers faced difficult physical and personal conditions to help. Not being ready to face such traumatic conditions has caused mental health issues for both unprepared spontaneous volunteers and first responders in similar disasters (Farfel et al., 2008; Sauer et al., 2014).

Disaster Volunteers. DOI: https://doi.org/10.1016/B978-0-12-813846-5.00006-X

Box 6.1 Best practices for volunteer management

In recent years, the U.S.-based National Voluntary Organizations Active in Disaster (NVOAD) has produced Points of Consensus to address volunteer management. As a central tenet, experienced NVOAD partners recommend that people should volunteer from within disaster experienced organizations in contrast to more spontaneously volunteering. Doing so links those wanting to serve with knowledgeable organizations that know how to meet local needs. With advance preparedness and planning, managers can transform spontaneous volunteering into a readily deployable group performing essential work safely and effectively. For more, visit https://www.nvoad.org/resource-center/member-resources/.

Why would people want to volunteer in such dire situations? People do so because they care, an expression of the ways in which we raise our children to be altruistic. Emergency managers and volunteer coordinators can anticipate that two kinds of altruistic volunteers may appear (Dynes, 1994). *Individual* altruism describes people who volunteer on a regular basis, in their own communities, when they perceive that needs exist. In contrast, *situational* altruism generates spontaneous volunteers: when people assume that unmet needs exist and they go in response to a disaster site (Dynes, 1994). Emergency and volunteer managers must be ready to take on this task and to use them effectively. This chapter addresses the benefits and challenges of spontaneous convergence in order to manage such volunteer energies well, along with guidance for how to retain such volunteers for the recovery time period (Cone, Weir, & Bogucki, 2003; Fritz & Mathewson, 1956; see also Box 6.1).

Benefits of spontaneous volunteers

Numerous benefits can accrue from spontaneous volunteers if managers prepare to use the energy they bring and the help they want to provide. Benefits include people being willing to travel significant distances, do difficult work, generate economic benefits, and buoy the spirits of those who survived, including emergency and volunteer managers.

To illustrate, consider that people may very well arrive from long distances at their own expense, resulting in considerable savings for local jurisdictions. The costs associated with travel to affected sites can be significant from car mileage to plane fares. Volunteers who shoulder this cost ease the burden of communities struggling to secure resources to fund basic commodities. Such financial independence can be particularly valuable to low-income areas. Besides the costs borne by spontaneous volunteers, experienced disaster organizations may also receive donations after a disaster, from people who believe in and support their efforts. Volunteers who want to go into a disaster site quickly should affiliate with such organizations in advance. Donated funds can be used to afford travel of both early-arriving and long-term volunteers, who would arrive when and where they are needed to assess the situation and set up long term projects.

Spontaneous and affiliated volunteers also willingly engage in truly dirty work, in hot, humid, or unpleasant conditions and then express an eagerness to do even more. If enough arrive and are managed well, their sheer numbers can help to pull off a significant amount of work that has to be done. To illustrate, one Midwestern U.S. city faced a daunting flood forecast that could inundate the entire area. Within hours, approximately 30,000 volunteers answered a call to help with sandbagging. Officials quickly set up stations, trained people to unload sand, and then fill, load, and place sandbags to reinforce a threatened levee system (Phillips, 1986). Their efforts proved successful, suggesting that such a convergent, unskilled labor force can help if coordinated well on a specific series of tasks. Imagine what could happen with an even higher degree of organization. In an impressive outpouring of rapid volunteering, residents around Salt Lake City built a sandbag levee to funnel snowmelt from the mountains through town and into a tributary. Arising out of the organizational structure offered by the Church of Latter Day Saints, volunteers also built temporary bridges over the makeshift river. Additional volunteers cooked for those involved in the levee effort. The city survived, because of a spontaneous but highly coordinated volunteer turnout (Quarantelli, 1994, for photos see https://www.ksl.com/article/41402975/looking-back-at-the-1983-flood-that-sent-a-river-through-downtown).

Another benefit comes from the economic value of volunteers. By collecting the number of people who show up and documenting what they do, the effort can be used as a matching portion for grant money in some locations. With advance preparation, managers can direct arriving volunteers to businesses that have been able to re-open. Volunteers will have to sleep, eat, shower, and buy supplies somewhere – and their money can help rejuvenate the damaged economy, re-open stores and restaurants, help residents earn money to pay insurance deductibles or repair homes, and restore the community's tax base. In Santa Cruz, California, post-earthquake efforts included signage directing visitors to their downtown stores and tourist areas. With highly visible signs reading "Open" and listing the businesses available, the area began to rebound economically. People who came to Vailankanni, India, after the tsunami also helped to restore economic vitality in a rapidly rebuilt commercial sector. People were able to feed their families because outsiders spent money locally, especially in an area struggling to also overcome loss of livelihoods in the fishing and agricultural sectors.

Another benefit comes from the psychological boost that survivors and affected residents experience when volunteers arrive. For those displaced and harmed by disaster, frustrated by governmental bureaucracy, or isolated due to location, the arrival of volunteers infuses a sense of relief to know that others care. Feeling their burden lifted, communities report a therapeutic effect that offers psychological benefits (Barton, 1970; Phillips, 2014, 2015). The arrival of this therapeutic community can be facilitated if managers anticipate SUVs. Thus, if local officials are ready – and if disaster volunteers wait to be called – their collective efforts can be integrated into a workable system that makes a huge difference.

Challenges of spontaneous volunteers

To be sure, SUVs also arrive with a number of challenges to be managed (Twigg & Mosel, 2017). After hurricane Andrew struck southern Florida in 1992, I watched as busloads of volunteers spilled out at what they perceived to be a main checkpoint near a large tent city. Eagerly stepping off the bus, the volunteers asked those they perceived to be in charge, "where do you need us?" Yet, the area was not yet ready for busloads of spontaneous volunteers to appear. As one problem, officials were still working up damage assessments to identify work areas. Within damaged areas lived specific populations needing help including elderly, people with disabilities, and single parents many of whom were not yet back from being evacuated. A lack of electricity meant that officials were communicating with residents via large balloons lofted over the area and color coded to indicate shelter, first aid, food, and water sites. Disaster organizations were cooking food for survivors at nearby schools and sending it out via nearly eighty emergency response vehicles that had arrived from across the U.S. Drivers faced challenges determining where to go because the high hurricane winds had destroyed street signs and traffic lights. It was hard to find a place to use a bathroom. The sweltering heat and humidity of southern Florida made the area difficult to endure. The volunteers who had disembarked left disappointed and upset.

Since Andrew, newer systems have evolved to link volunteers with needs, though sometimes these emerge outside the knowledge of emergency and volunteer managers, such as with spontaneous social media efforts (more on this later). So, even decades after Andrew, personal convergence and spontaneous volunteering remains problematic. Managers may not know who has arrived or how many are coming or for what reasons. Ideally, we want to place the right volunteers in the right places at the right time – with the right skills suited for the job, places, and people and to do so in a timely manner.

Volunteers will also want and need local guidance to navigate what can be a disturbing, unsafe, and complex post-disaster environment.

While volunteers ideally arrive ready to work, many will be surprised by what a disaster can generate and lack proper protective equipment, tools, gloves, clothing, or footwear. They may not realize they cannot use or recharge cell phones or that vehicle gas might not be available. The disaster may have damaged places to stay or survivors may need available units for their own temporary housing. Sanitation facilities, from toilets to showers, might be unavailable or, like with earthquakes, the disaster may have compromised the entire water and wastewater system. Volunteers may thus need food, water, first aid, and potential advanced medical support should a serious emergency occur. In Joplin, the tornado destroyed the hospital, limiting abilities to handle volunteer injuries or emergencies. In the aftermath of a disaster, we do not want more injuries or harm to occur. If they do, we do want people cared for quickly.

Accordingly, some degree of supervision and safety training should be required. Buildings might appear safe, for example, but structural collapse could be imminent. Power lines could be down but not off, creating a life-threatening situation for volunteers moving into an area. Avoiding that "debris pile" selfie photo would be a good idea, especially with unknown chemicals, nails, splintered wood, glass, and voids within the debris. Trees could also be unstable and debris could be hanging precariously from limbs. Some hazards, like flooding, will alter local environmental patterns, with new threats from hazardous materials that have spread into yards, walkways, and open spaces. Post-Katrina, studies surfaced multiple concerns about sites like "service stations, pest control businesses, and dry cleaners" as well as "chemical plants, petroleum refining facilities, and contaminated sites, including Superfund sites" that could have spread into areas where volunteers would work (Reible, Haas, Pardue & Walsh, p. 6). Coupled with other contaminants including household chemicals, lumber soaked with creosote, human and animal waste, mold, and pre-Katrina contamination, people worried about exposure. Nonetheless, hundreds of thousands of volunteers came, and usually with minimal safety protection or follow-up assessments to determine any short or long-term health care impacts.

It can be particularly hard to monitor and follow spontaneous volunteers for such follow-up. Increasingly, managers are using some type of registration and identification system. Well-prepared communities identify entry points for spontaneous volunteers, establish disaster perimeters, and require badging. Identification badging provides several degrees of added security (see also Chapter 5). First, managers will know who came and where they came from. Such information can help to amass a database for future use during the recovery time period. Second, badging also helps first responders when volunteers become injured or ill and can be linked to emergency contact information or include medical conditions. Second, a database allows for potential follow-up, in case an exposure becomes apparent or a contagious disease develops. As managers deploy volunteers, a well-prepared effort can map the volunteers in real time to monitor work efforts and insure safety. For example, if inclement weather develops, managers can warn those on work sites and direct them to shelters. In Japan, those near Fukushima Daiichi nuclear plant accident continue to be monitored including workers who bravely volunteered to stay at the site to address the accident. It is expected that a number will develop cancer as a consequence of their exposure (Ten Hoeve & Jacobson, 2012).

Another challenge that may need to be addressed concerns unanticipated impacts on local economic interests. Spontaneous volunteers, while well-meaning, may displace local workers. As a rule, volunteer labor should support rather than supplant locally employed people trying to rebuild their lives. Yet problems erupt in many disasters where local people lose work opportunities. After the Indian Ocean tsunami, non-governmental organizations hired people and used volunteers to rebuild homes not realizing they were displacing women from traditional roles in making decisions about home construction (Barenstein, 2006). That displacement costs people money and influence, as well as the ability to self-govern their own recovery.

Managers will clearly have to walk a bit of a tightrope in order to manage volunteers eager to serve. Careful diplomacy in messaging and managing spontaneous volunteers will be needed. Otherwise, volunteers may leave, unhappy they could not help and share their feelings on social and traditional media – which could discourage future volunteers. In short, if managers anticipate the outpouring of generosity that will come after a disaster and be ready to direct and manage it appropriately their community will benefit. Managers should be ready to launch a volunteer management plan immediately including non-deployment messaging, online or app-based registration, and proper use of skilled and unskilled help. Given that volunteers now interact in an online and social media environment, managers have to be there as well.

Electronic convergence

The last decade has revealed the power, potential, and challenges associated with volunteering via the Internet, particularly in the dynamic evolution of online platforms, apps, and social media. Electronic tools have generated widespread and easily shared coverage of disasters, motivated large volunteer turnouts, created new tools to crowdsource solutions to disaster needs, and surfaced additional benefits and problems with spontaneous volunteers. It is clear that emergency and volunteer managers need to be smart and to monitor electronic means to communicate because people are using them to converge electronically. In this section, we look at a few key themes that span electronic convergence from information spread to remote volunteering.

To start, social media fuels information dissemination (correct or not) and electronic convergence. In the 2014 Australian bushfires, 50% of all Twitter messages came from the general public (Abedin & Babar, 2018). The other half came from emergency response organizations and related agencies. The result? An extensive number of tweets and retweets poured out and created significant challenges. In the majority of the Australian bushfire tweets, the public failed to geo-locate information making it difficult for responders to know where to send help or volunteers. As a consequence, response organizations had to sort through a large volume of information and determine what was accurate and actionable (Abedin & Babar, 2018). Misinformation may erupt as well, from the unintentional to the intentional use of social media by terrorists, such as in the 2008 attack on Mumbai, India (Oh, Agrawal, & Rao, 2011). Thus, the first task of addressing electronic convergence comes from dealing with the tsunami of information people send out, share and repeat, thinking that they are being helpful. Retweeting or reposting is not necessarily helpful without actionable information.

An historic event may have changed that, albeit not without a lot of problem-solving around it. After the Haiti earthquake in 2010, volunteers located around the planet leveraged a Kenyan-made platform called Ushahidi (Heinzelman & Waters, 2010; Hughes & Tapia, 2015). Using this open-source software, volunteers analyzed social media reports and mapped emergency relief situations. A number of challenges impeded the volunteers, which they worked to overcome. To start, volunteers lacked accurate maps of Haiti. To deal with that problem, volunteers relied on other open source tools like Google Earth and OpenStreetMap. Language differences also impeded the effort, as Haitians often reported needs in Creole. Stanford University worked with Ushahidi to design a new system that would remotely translate and geo-code messages. Their efforts created a system that accomplished this critical task within ten minutes from receipt to posting. They then expanded their efforts further by collaborating with several companies and the U.S. State Department. The crowdsourced, international effort translated over 25,000 messages by creating and mapping in excess of 3500 actionable items (Heinzelman & Waters, 2010). A similar result occurred after Typhoon Haiyan struck the Philippines in 2013. At least 1600 online volunteers accessed OpenStreetMaps to geo-locate needs (Westrope, Banick, & Levine, 2014).

A final challenge emerged in Haiti when relief requests and reports dropped off after the United Nations announced that the emergency time period had ended. Yet, the recovery lingered for nearly ten years, with volunteers needed well beyond the emergency time period. A similar drop-off happened after Hurricane Sandy in the U.S. (see Lachlan, Spence, Lin, & Del Greco, 2014). Since Haiti, electronic convergence, crowdsourcing, and digital volunteerism has become even more far-reaching (Ludwig, Kotthaus, Reuter, van Dongen, & Pipek, 2017; Starbird, 2011; Starbird & Palen, 2011). The use of hashtags (for example, #disaster, #Maria, #PuertoRico) has enabled self-identified groups to find information faster and to connect (Starbird, 2011). However, social media does not seem to be particularly helpful yet with generating long-term recovery volunteering.

Some efforts have become more institutionalized, by using a hybrid approach. The Netherlands Red Cross created Ready2Help to funnel convergence. The platform works by pre-registering volunteers who would be reached via traditional means like email. To test the system, the Red Cross conducted a field exercise and tried to connect volunteers to incoming tweets. However, the tweets became too voluminous to handle in the exercise, similar to what happened in an Australian event (Schmidt, Wolbers, Ferguson, & Boersma, 2017). Nonetheless, the Netherlands Red Cross activated their system in 2015 during a refugee crisis, demonstrating an ability to manage public convergence during a slower-moving event. Similar efforts have been adopted in other areas. Austria established Team Österreich as far back as 2007 while Czechoslovakia uses TEAM MORAWA *if* volunteers register (Neubauer et al., 2013).

Thus, while challenges clearly exist with electronic convergence, the potential also exists to crowdsource, direct, and leverage spontaneous volunteers. Accordingly, emergency and volunteer managers would be well-advised to have a high-tech team involved in their volunteer management plans. Due to the continual and often rapid evolution of these tools, that team must be capable and current to know which tools and platforms to use and who uses them. With the youngest generations switching platforms frequently, it may be challenging to reach, educate, register, and deploy younger volunteers. Devices also change quickly, although the interoperability of websites and social media platforms on those devices has become increasingly sophisticated. Social media companies also change their platforms and algorithms in ways that alter messaging from who sees what and when to how many characters may be used. Still, people are using small screen phones as a primary device for all forms of communication including in remote areas worldwide.

Challenges do exist with such technology. As a volunteer managing a triage reception area for an animal response team, the power cords that I duct-taped to my table proved invaluable to my teammates. After the Haiti earthquake, survivors set up charging stations in the tent city encampments. Another challenge could occur with cellphone disruptions caused by disasters such as tornados that take out cellphone towers, geomagnetic storms that disrupt connections, or cyberattacks that produce denials of service. Nonetheless, a majority of the public uses social media daily (Merchant, Elmer, & Lurie, 2011) and solar-powered chargers can be used. Emergency and volunteer managers need to remain updated and to leverage use of these platforms for today's "wired" volunteers.

Managing spontaneous volunteers

Within a few days after a 1999 tornado outbreak in Oklahoma, Bridge Creek schools had used their location to set up radio communications, donations reception and dispersal, food distribution, and volunteer coordination. They had supported emergency rescue efforts and invited the federal government to use their facilities for the recovery. Teachers had organized donations and set bags of supplies on each student's desk to await their return to school. Yet, only one staff member had prior disaster training before the EF5 tornado bore down on the area. Imagine what could happen with advance planning.

Managers can anticipate the arrival of spontaneous volunteers by planning with other agencies and organizations, such as volunteer centers, health care units, faith communities, utility companies, animal rescues, social service agencies, and civic clubs. A volunteer management plan, as generally described in Chapter 5, should be seen as a traditional and necessary part of any community's general disaster recovery plan and should be systematic (Fernandez, 2006, for more guidance see Box 6.2). Many existing volunteer

Box 6.2 Resources for managing spontaneous volunteers

A number of sites offer general guidance and tools for managing volunteers in general and spontaneous volunteers in particular. For more information, visit:

The Corporation for National and Community Service

For downloadable guidance on Volunteer Reception Centers, risk management, and managing spontaneous volunteers, go to:

https://www.nationalservice.gov/sites/default/files/olc/moodle/ds_managing_volunteers_in_time/viewbfff.html?id = 3198&chap terid = 1924

Federal Emergency Management Agency

This U.S. agency offers a pdf titled "Managing Spontaneous Volunteers" that is most easily found through a search engine. In addition, FEMA offers free online courses including IS-244b. Developing and Managing Volunteers at https://training.fema.gov/is/crslist.aspx

City of Los Angeles Community Emergency Response Team

Visit this site to see how Los Angeles manages spontaneous volunteers through recommending pre-disaster training, https://www.cert-la.com/cert-training-education/spontaneous-volunteers/.

Guidelines for Proper Debris Management and Removal

Spontaneous volunteers will often be involved in dealing with disaster debris which must be disposed of properly. For guidance, visit https://www.fema.gov/disposing-debris-removing-hazardous-waste.

All sites last accessed August 21, 2019.

management plans focus heavily on the need to manage spontaneous volunteers. Elements of such a plan specifically for spontaneous volunteers should include:

- Designating a lead agency to coordinate the influx and pre-establish SUV-focused:
  - Connections between organizations that support key needs such as health care and first aid, food and hydration services, parking, and toilets.
  - MOUs for basic amenities including sanitation, hydration, and first aid support.
  - Ties to damage and needs assessment teams to secure actionable information, identify appropriate work locations, and deploy incoming volunteers effectively.
  - Pre-arrival messaging for multiple platforms and media, preferably with the support of a team dedicated to electronic convergence.
  - Pre-identification of entry reception site(s) for incoming volunteers that will be widely publicized. Several sites should be identified because disasters can destroy planned sites or access to the site. Stations for intake, training, safety education, local information, transportation, first aid, food and hydration should be designated inside the site.
  - Forms (hard copy and electronic if the power/generator is on) for capturing volunteer information including identity information for badges, emergency contacts, and skill sets. See Box 6.3 for an example usable for spontaneous and affiliated volunteers.
  - An established means for pre-identified people with experience to train quickly the arriving volunteers, typically around the most likely disasters to happen in the area.
  - A cache of tools and protective equipment as well as a set of procedures for checking out and insuring the return of said items.

Box 6.3 Intake of spontaneous volunteers

What would a manager want to know about the volunteers coming in? Volunteer Florida, where extreme weather routinely happens, is ready for spontaneous volunteers. They have designed a registration form that captures an extensive amount of information. Their efforts include demographic backgrounds of volunteers and emergency contacts as well as health limitations. For skill sets, they ask about medical, veterinarian, communications, languages, office skills, structural and damage assessment skills, transportation skills, skilled labor and equipment operation and general services from food to elder and child are, social work, traffic control, animal care, and disability support. Their registration form also includes a release of liability statement and a means to record and verify credentials including background checks. For more information, visit www.volunteerflorida.org.

- Written and electronic instructions on work site safety coupled with training on how to volunteer safely as well as where and how to find emergency help should injuries occur. Multiple languages should be offered including sign language and sites should be accessible. Everyone will want to volunteer and managers should find ways to involve them.
- A means to conduct evaluations (e.g., online surveys) in order to capture lessons learned for the next time. Area universities might be able to help with such surveys in exchange for access to anonymized data. A link to an online survey could be sent out a few days after the volunteers have departed so that they will have had time to reflect on their experience.
- Both informal and formal means to recognize and thank volunteers. Social media can be particularly helpful with this task.
- Training on the plan, preferably via a practical exercise, and then revision of the plan based on insights gained through the exercise.
- Considering the plan to be a living document that can and should be adapted annually.

Next, local people and agencies skilled in volunteer work can assist with creating a Volunteer Reception Center (see Chapter 5). For SUVs, special consideration should be given to:

- Coordinating with the local emergency management agency to confirm that volunteers are needed and, if so, where they should be deployed.
- Monitoring online websites, social media, and related platforms immediately as it may be the best source of information prior to the arrival of SUVs.
- Messaging to potential volunteers, in coordination with the local emergency management agency. Messages should be consistent and be issued simultaneously and widely with a range of partners (via traditional and social media, with a unique #hashtag). Messages should include clear instructions on where to go and what to do including: "Do not deploy until notified." Multiple languages, including sign language, should be used because everyone will want to help. Encourage spontaneous volunteers to arrive self-contained with all they need (and to identify any open stores, restaurants, et al.) and to use an online registry to volunteer. By educating people who want help, spontaneous volunteers may understand why you want them to wait. Always be diplomatic, because they do mean well and may be able to help at the right place and time.
- Insuring that a reception site is accessible due to the disaster and for volunteers arriving with accessibility needs because everyone can help.
- Electronic convergence promotions, with an eye toward correcting misinformation.

- Operating the intake/registration site carefully, so that local managers know who came and where they are. Volunteers from various groups may become separated into various work sites, so being able to locate and/or reunite them successfully is helpful – especially when cell phones do not work properly or cannot be re-energized. Identification of those spontaneous volunteers also helps if an emergency occurs. Barcoded bracelets may be useful to track entry and egress from a disaster site for safety purposes. Badges are helpful as well as the ability to conduct background checks as needed.
- Creating a training area as needed for specific issues from sandbagging to sorting debris for later appropriate disposal. Given the timeline in which spontaneous volunteers arrive, guidelines for the complexities of debris sorting should be made widely available. This will help to minimize risk to volunteers such as exposure to hazardous materials and to avoid re-sorting. Voluntary organizations do report that spontaneous volunteers have been injured, from scrapes and cuts to insect bites, respiratory issues, post-traumatic stress and death (Aakko, Weed, Konrad, & Wiesman, 2008; Debchoudhury et al., 2011; Sauer et al., 2014; Swygard & Stafford, 2009; Thormar et al., 2010). Accordingly, always emphasize safety at all times.
- Establishing perimeters and entry points for highly damaged areas. People naturally want to see what happened but access is advisable to reduce injuries and maximize incoming volunteer energy.
- Transporting volunteers into an area of need may be wise given that debris may linger and damage personal vehicles. In addition, fewer vehicles in a given area means better access for emergency vehicles and fewer tow trucks needed for disabled volunteer vehicles.
- Determining how to give priority to specific areas, such as a retirement facility, a school, or a faith-based location. By putting the efforts of a large group toward such sites, a foundation can be laid to help the community leap forward more quickly. Further, identifying the needs of people who cannot recover without help, such as elderly residents, helps to get them back to normal faster.
- Supervising spontaneous volunteers including work site and roving supervisors. Difficult conditions can tax inexperienced volunteers and a moment of inattention could make the difference with injuries. Construction companies or home supply stores may be able to provide such supervisors during the crisis, when their employees are unable to work.
- Monitoring worksite progress should start as soon as possible in order to deploy additional arriving volunteers. Monitoring could be as simple as sticky notes on the wall or a whiteboard that is updated regularly. Given electronic volunteers, it may be possible to map volunteer work sites and progress.

- Evaluating and following up. Many spontaneous volunteers come for the day and then leave. Having seen the damage, though, may inspire them to return at a future date for savvy managers who capture their information and invite them back. Doing so requires committing to finding out how their experience went and the means through which volunteers could be contacted again. Managers can use the information to improve the plan for the next time.

Retaining spontaneous volunteers

Keeping spontaneous volunteers for the long-term recovery represents an essential component of volunteer management. As part of the pre-disaster volunteer planning (or after the disaster as is often necessary), managers should try to create databases and messaging systems to keep volunteers informed about ongoing needs and efforts. Some ideas to do that include:

- Setting up social media accounts to keep volunteers up to date on what is happening and create events to spur recovery. Let them know how, when, and where they can help, link to experienced organizations, and be of value to your community.
- Creating volunteer registries where people can request help and the organization can match volunteers. Such registries, usually web-based, may emanate from existing volunteer centers or one that steps up after the disaster. Digital volunteers may be able to help with this.

After the Joplin, Missouri tornado, many animals became lost. The local Joplin Humane Society provided emergency shelter, veterinary care, food, and grooming. Later, pets that never found their families were placed for adoption. Because the Humane Society had an established social media presence and web site, they were able to keep people informed of an impending adoption event and worked with the American Society for the Prevention of Cruelty to Animals to make it happen. Prospective adopters came from dozens of states to adopt hundreds of displaced animals, through a carefully coordinated process to insure they found their forever homes. Pets went to new homes of people who volunteered to save them, providing comfort and security after a deeply traumatizing disaster. Just by keeping people informed, the Humane Society effectively brought caring volunteers to their community at the just right time (see also Box 6.4).

Part of retaining volunteers means recognizing what they bring into your community. Their arrival means that not only has help arrived – but so has hope. So – recognize and reward those volunteers for making a difference whether they arrive spontaneously or via affiliated organizations. One meaningful way to recognize volunteers is to link them with survivors. Though care must be taken to not add to the burden of survivors, encounters can

Box 6.4 And in the Debris, they Found. . .koi.

Volunteers have a huge heart for service, and can have an eye for what others miss. In 2013, I volunteered for a pre-established and pre-trained Oklahoma animal response team helping save pets and livestock after an EF5 tornado. Spontaneous volunteers had arrived from many locations and poured into neighborhoods helping to sort through and remove debris. Within those neighborhoods, people's pets could still be saved. Volunteers called or came in with reports of loose animals that our teams went out to find or trap. As volunteers sorted through debris, they also found animals injured but alive and brought them to our triage unit where veterinarians waited or team members transported them to animal hospitals. Uninjured animals went to a pet shelter staffed heavily by local volunteers who cleaned, fed, and cuddled with traumatized animals. At a fairgrounds, people capable of working with large animals tended to horses, cows, and goats.

In such a unit, the incident command system (ICS) is used to establish who is in charge and who does what. My job became establishing a database for found pets and helping people and animals as they came into triage and moved on to veterinary care. I rarely interacted with the Incident Commander, who was quite busy managing the situation. Then, a woman and her teenage children came in. They were wearing t-shirts from a voluntary organization and had arrived to ask for help. In cleaning debris from someone's home, they had discovered a pond of koi fish facing immediate peril. Debris had landed in their pond, which was rapidly dehydrating in the high heat and humidity of Oklahoma in May. The koi were beginning to show some signs of distress with limited space and diminishing pond water. Our unit had just dispatched a number of vehicles to rescue a stranded pygmy goat, so I felt confident we could do something. I noticed that the Incident Commander was standing by himself and approached him. "Sir," I said, "we have a rescue effort involving koi that needs to happen." He looked at me, "koi?" he said. "Yes, sir, yes – koi." I pointed out the volunteers who had come in to report the situation and in an opportune moment, the teenaged girls waved at him hopefully. He sighed, looked back at the volunteers, said "give me a few minutes," and then pulled out his phone. Shortly, he had secured beverage coolers and arranged the koi rescue to an available pond. The koi, for me, came to represent the best of spontaneous volunteers who discern unexpected needs and work through the process of arranging for help alongside experienced volunteers trained and affiliated within an organized response system. I suspect the koi appreciated it as well.

make everyone feel better. Those encounters might occur over a joint meal at a local place of worship or community center. Meaningful interactions might happen on a disaster site when survivors and volunteers work alongside – or they could take place at the end of a volunteer shift when a group of survivors collectively thank those who helped. That human connection, built meaningfully between hurt and hope, matters. It makes local residents feel uplifted and can inspire volunteers to serve again, whether in another damaged community or elsewhere. It can make survivors feel less obligated by being able to issue a heartfelt "thank you." Thus, volunteers have the potential to generate a "therapeutic community" that makes people feel better in the face of a daunting recovery (Barton, 1970; Phillips, 2014, 2015). Survivors' futures look less bleak and the frustrations they may feel with formal response efforts may be defused.

Volunteer and emergency managers can also organize local thank you events and public notices while volunteers are still in the community, perhaps through the use of social media, electronic billboards, handmade or professional signs, and news stories. Locally operating businesses may want to participate in appreciating volunteers, with donated resources, signs in their windows, or discounts. Local leaders can work with worship locations to host celebrations or appreciation meals or school children can write thank you notes to be given out at the volunteer reception center. Teenagers could do live reporting via social media on volunteer sites. Managers can also take group photos and spread them through traditional and social media.

Communities can also be good hosts and offer guidance on where to find housing, food, supplies, medical care, and tools. Rewards can also include offering tourism opportunities, perhaps free or with discounts, to spur the local economy back into operation. There is a lot that can be done to show appreciation and to lure volunteers back for the long-term recovery. Communities damaged by disasters are going to need them.

Conclusion

Volunteers will come, so managers need to be ready for them whether they arrive in person or through electronic convergence. Being ready starts with having a well-developed volunteer management plan in place and being ready to direct the altruistic energies of those heading to help. Managers need to be prepared to educate spontaneous volunteers about when and where to come and then to use those volunteers once they show up. Being ready at a set volunteer reception center also enables managers to leverage the energy of SUVs. Moderating electronic convergence can both manage incoming SUVs and develop potentially useful collaborations. By being ready for the arrival of well-meaning spontaneous volunteers, managers can address specific needs that have emerged through damage and needs assessments. Onsite management will necessarily including rapid training for specific tasks along with safety training and appropriate supervising. Being prepared for all aspects of SUV arrivals, from start to finish, matters. They do mean well, so be sure to find ways to recognize and thank them.

References

Aakko, E., Weed, N., Konrad, R., & Wiesman, J. (2008). Rethinking volunteer management using a centralized volunteer staging and training area. *Disaster Medicine and Public Health Preparedness*, *2*(2), 127–129. Available from https://doi.org/10.1097/DMP.0b013e31816476a2.

Abedin, B., & Babar, A. (2018). Institutional vs. non-institutional use of social media during emergency response: A case of Twitter in 2017 Australian bush fire. *Information Systems Frontiers*, *20*(4), 729–740.

Barenstein, J. (2006). Challenges and risks in post-tsunami housing reconstruction in Tamil Nadu. *Humanitarian Exchange*, *33*, 39–40.

Barton, A. (1970). *Communities in disaster: A sociological analysis.* Garden City, NY: Anchor Books.

Cone, D. C., Weir, S. D., & Bogucki, S. (2003). Convergent volunteerism. *Annals of Emergency Medicine, 41*(4), 457–462.

Debchoudhury, I., Welch, A. E., Fairclough, M. A., Cone, J. E., Brackbill, R. M., Stellman, S. D., & Farfel, M. R. (2011). Comparison of health outcomes among affiliated and lay disaster volunteers enrolled in the World Trade Center Health Registry. *Preventive Medicine, 53*, 359–363.

Dynes, R. R. (1994). *Situational altruism: Toward an explanation of pathologies in disaster assistance.* Preliminary Paper #201. Newark, DE: University of Delaware, Disaster Research Center.

Farfel, M., DiGrande, L., Brackbill, R., et al. (2008). An overview of 9/11 experiences and respiratory and mental health conditions among World Trade Center Health Registry enrollees. *Journal of Urban Health, 85*(6), 880–909.

Fernandez, L. S. (2006). Strategies for managing volunteers during incident response: A systems approach. *Homeland Security Affairs, 2*(3), 1–15.

Fritz, C. E., & Mathewson, J. H. (1956). *Convergence behavior: A disaster control problem. Special report prepared for the Committee on Disaster Studies.* Washington D.C: National Academy of Sciences, National Research Council.

Heinzelman, J., & Waters, C. (2010). *Crowdsourcing crisis information in disaster-affected Haiti.* Washington D.C.: United States Institute of Peace.

Hughes, A., & Tapia, A. (2015). Social media in crisis: When professional volunteers meet digital volunteers. *Journal of Homeland Security and Emergency Management, 12*(3). Available from https://doi.org/10.1515/jhsem-2014-0080.

Lachlan, K. A., Spence, P., Lin, X., & Del Greco, M. (2014). Screaming into the wind: Examining the volume and content of tweets associated with Hurricane Sandy. *Communication Studies, 65*(5), 500–518.

Ludwig, T., Kotthaus, C., Reuter, C., van Dongen, S., & Pipek, V. (2017). Situated crowdsourcing during disasters: Managing the tasks of spontaneous volunteers through public displays. *International Journal of Human-Computer Studies, 102*, 103–121.

Macintyre, A. G., Barbera, J. A., & Smith, E. R. (2006). Surviving collapsed structure entrapment after earthquakes: A time-to-rescue analysis. *Prehospital and Disaster Medicine, 21*(1), 4–17.

Merchant, R. M., Elmer, S., & Lurie, N. (2011). Integrating social media into emergency-preparedness efforts. *New England Journal of Medicine, 365*(4), 289–291.

Neubauer, G., A. Nowak, B. Jager, C. Kloyber, C. Flachberger, G. Foitik, & G. Schimak. (2013). Crowdtasking – a new concept for volunteer management in disaster relief. International Federation for Information Processing, ISESS 2013, AICT 413, pp. 345–356.

Noji, E., Armenian, H. K., & Oganessian, A. (1993). Issue of rescue and medical care following the 1988 Armenian earthquake. *International Journal of Epidemiology, 22*(6), 1070–1076.

Oh, O., Agrawal, M., & Rao, R. (2011). Information control and terrorists: Tracking the Mumbai terrorist attack on Twitter. *Information Systems Frontiers, 13*(1), 33–43.

Phillips, B. (2014). *Mennonite Disaster Service: Building a therapeutic community after the Gulf Coast storms.* Lanham, MD: Lexington Books.

Phillips, B. (2015). Therapeutic communities in the context of disaster. In A. Collins (Ed.), *Hazards, risks, and disasters in society* (pp. 353–371). Elsevier.

Phillips, B. D. (1986). Mass media roles in disaster threat situations: Recruitment and portrayal of the emergency untrained volunteer. *International Journal of Mass Emergencies and Disasters, 4*, 7–26.

Quarantelli, E. L. (1994). Emergent behavior and groups in the crisis time period of disasters. Preliminary Paper #206. Newark, DE: University of Delaware, Disaster Research Center.

Sauer, L. M., Catlett, C., Tosatto, R., & Kirsch, T. D. (2014). The utility of and risks associated with the use of spontaneous volunteers in disaster response: A survey. *Disaster Medicine and Public Health Preparedness*, *8*(1), 65–69.

Schmidt, A., Wolbers, J., Ferguson, J., & Boersma, K. (2017). Are you Ready2Help? Conceptualizing the management of online and onside volunteer convergence. *Journal of Contingencies and Crisis Management*, *26*(3), 338–349.

Starbird, K. (2011). Digital volunteerism during disaster. *Crowdsourcing Information Processing*. Available from https://www.humancomputation.com/crowdcamp/chi2011/papers/starbird.pdf, Accessed 16.05.19.

Starbird, K. & L. Palen. (2011). Voluntweeters: Self-organizing by digital volunteers in times of crisis. CHI 2011. Available from https://cmci.colorado.edu/~palen/voluntweetersStarbirdPalen.pdf, Accessed 16.05.2019.

Swygard, H., & Stafford, R. E. (2009). Effects on health of volunteers deployed during a disaster. *American Surgeon*, *75*, 747–753.

Ten Hoeve, J., & Jacobson, M. (2012). Worldwide effects of the Fukushima Daiichi nuclear accident. *Energy & Environmental Science*. Available from https://doi.org/10.1039/c2ee22019a.

Thormar, S. B., Gersons, B., Juen, B., Marschang, A., Djakababa, M. N., & Olff, M. (2010). The mental health impact of volunteering in a disaster setting, a review. *Journal of Nervous Mental Disorders*, *198*, 529–538.

Twigg, J., & Mosel, I. (2017). Emergent groups and spontaneous volunteers in urban disaster response. *Environment & Urbanization*, *29*(2), 443–458.

Westrope, C., Banick, R., & Levine, M. (2014). Groundtruthing OpenStreetMap building damage assessment. *Procedia Engineering*, *78*, 29–39.

Further reading

Phillips, B., Neal, D. M., Wikle, T., Subanthore, A., & Hyrapiet, S. (2008). Mass fatality management after the Indian Ocean tsunami. *Disaster Prevention and Management*, *17*(5), 681–697.

Reible, D., Haas, C., Pardue, J., & Walsh, W. (2006). Toxic and contaminant concerns generated by hurricane Katrina. *The Bridge Linking Engineering and Society*, *36*(1), 5–13.

Siemen, C., dos Santos Rocha, R., Van den Berg, R. P., Hellingrath, B., Porto de Albuquerque, J. (2017). Collaboration among humanitarian relief organizations and volunteer technical communities: Identifying research opportunities and challenges through a systematic literature review. ISCRAM Conference, Albi, France

CHAPTER 7

Affiliated disaster volunteers

Why join an experienced disaster organization?

Disaster volunteers help in many ways, working to address unmet and sometimes unique needs for humans and animals as well as built and natural environments. When volunteers help disaster-stricken communities in an organized and concentrated manner, they can generate an incredible impact. If those efforts emanate through coordinated disaster organizations, their collective and often collaborative efforts can restore individual lives, entire neighborhoods and communities. To inspire affiliation with organized disaster relief volunteering, this chapter will review the types of organizations that function in a disaster setting, why volunteers should affiliate with them, and the powerful effects they can have when they collaborate. This section will be followed by a consideration of the kinds of volunteers that provide unskilled, skilled, and professional labor through disaster organizations. Opportunities abound to become affiliated with an organization suited to one's interests or to build connections to them as an emergency or volunteer manager.

Organizational types in disasters

As mentioned in Chapter 3, a classic typology can shed light into the kinds of organizations that onboard volunteers when disaster strikes (see Figure 3.6; Dynes, 1974). Two of those types have historically held disaster missions and are called established and expanding organizations. *Established* organizations include those with disaster experience, and existing structures and tasks such as CARE, Doctors without Borders, or emergency management agencies. *Expanding* organizations like the Red Cross/Red Crescent offer considerable experience through both paid and unpaid workers thus while shouldering familiar tasks may take on a larger or new structure. Two other kinds of organizations, emergent and extending, tend to appear as a result of an actual or potential disaster. *Emergent* organizations appear around a threat or because of unmet needs and create a brand new organizational structure and set of tasks to take on. An example would be a

Disaster Volunteers. DOI: https://doi.org/10.1016/B978-0-12-813846-5.00007-1

woman-centric organization that creates new employment opportunities post-disaster (Lund & Vaux, 2009). *Extending* organizations, with familiar structures, operate by offering resources like construction companies that remove debris or elevate homes out of floodplains as new tasks.

While each of these four types of organizations may include traditional, paid employees, the heart of the typology inherently embraces volunteerism and an altruistic ethic. For example, established organizations such as law enforcement or firefighters include volunteer groups like Volunteers in Police Service. Emergent organizations nearly always develop out of volunteer time and commitment and are often community or interest based. Extending organizations will send out teams or assets without compensation and organize philanthropic efforts to help people and places harmed by disasters. Expanding organizations bring in volunteers to augment their existing staff like many faith-based organizations do after a disaster or some rely entirely on volunteer commitment. This wide range of organization types active in disaster means that many opportunities to volunteer will be available. In contrast to spontaneous volunteers, those who work within disaster organizations and efforts that extend critical resources can pool their efforts.

Why be affiliated?

Why should volunteers join organized efforts? Disaster organizations know what they will face based on prior experience not only within their organization, but with survivors, officials, situations, and with similar organizations thus providing a context in which volunteers can be prepared to serve (Phillips, Mwarumba, & Wagner, 2012). Experienced organizations also prepare their volunteers for specific, familiar tasks and usually offer supervision, training, and safety oversight. They will also understand how to arrange for travel, housing, food, showers, supervision, and interaction with survivors. Their prior disaster experience enables them to know how to work with community officials on reconstruction requirements, building codes, and permits. Their prior work also provides a stewardship framework for using donated funds and resources wisely to avoid overlap with other organizations. Disaster organizations also network and collaborate with each other in advance, providing an intentional and purposeful effort to address a wide array of needs. All of these benefits mean that volunteers will be able to engage in productive, meaningful work that leverages their abilities and energies more effectively.

Do affiliated organizations come with challenges as well as benefits? They do, and perhaps the first one is funding. Most disaster organizations rely on donations, grants, and fundraising to sustain their usually minimal staff, infrastructure, and operational budgets. Despite relying on volunteers, funds will be needed for tools, travel costs, staffing/supervision, communication, computers, food, lodging, and resources to meet disaster needs (see Photos 7.1 and 7.2). Most disaster organizations rely on a small number of paid personnel or may use unpaid volunteers who function like paid staff. Having staff, including supervisory personnel, is necessary to locate volunteer efforts, coordinate with local agencies and officials, make sure that volunteer efforts meet codes or other requirements, and provide for the safety of volunteers. In addition to locating volunteer efforts, staff can also scout ahead for housing, arrange for utilities, and set up a means to feed helpers. Staff would also be tasked with financial stewardship of funds used for their work as well as transparency and accountability to donors and homeowners.

A second challenge often comes from managing an influx of volunteer interests after a major disaster, because most people want to help right after an event. Staff can manage those interests and focus people into teams that arrive with the right skills at the right time in the right places. Nonetheless, it can be challenging for disaster organizations to encourage people to "wait." By collaborating among disaster organizations, more people can be helped and donations as well as volunteer effort can be extended further and more effectively. One strategy that has been used in the U.S. has come from a movement that emerged in the 1960s to avoid duplication and encourage collaboration, the Voluntary Organizations Active in Disaster (VOAD) movement as first described briefly in Chapter 1.

PHOTO 7.1
Housing needs for volunteers should provide comfortable, clean, and restful environments.

PHOTO 7.2
Volunteers will need tools to work on long-term recovery projects.

The National VOAD movement

Over the past five decades disaster organizations have realized they need to coordinate to address massive needs, reduce overlap, and use their resources wisely. In the U.S., that collective realization merged into an umbrella-type organization called the National Voluntary Organization Active in Disaster (NVOAD). Over time, NVOAD spread as a movement, forming VOAD units at the local and state levels in the U.S. and serving as a model for best practices. Today, NVOAD has worked through decades of mindful consideration to create four central tenets for disaster organizations (all available at www.nvoad.org, last accessed July 11, 2019). As a set of principles to inform their efforts, members commit to the 4 C's: cooperation, communication, coordination, and collaboration.

The Four C's play out well in a disaster. A typical post-disaster scenario would likely see a number of organizations convening in a common location like a community center, worship location, or an office building. Someone, perhaps a local VOAD chair, emergency manager, or Voluntary Agency Liaison (see Chapter 1) would lead the organizations present to discuss apparent, emerging, and anticipated needs. Given their collective experience, organizations will likely volunteer for various tasks or locations suitable to their capabilities, volunteer base, and missions. Many member organizations offer specialized missions for a range of disaster tasks: case management, child care, clean up, donations management, mass feeding, medical care, animal support, shelters, repairs and rebuilding. Knowing the full array of affiliated organizations and their missions can offer managers a set of resources and provide volunteers with opportunities specific to their interests and abilities.

Building on the 4 C's, members have worked diligently the past ten years to create related Points of Consensus (POCs). Their POC for Volunteer Management says that volunteers "when properly coordinated, make up an essential part of the human resources needed to respond to disasters of all magnitudes" (NVOAD, 2011). The volunteer management POC sets out both volunteer rights and volunteer responsibilities. Volunteers should have the right to be treated with respect, to be valued, to participate in an "organized, structured system that matches skills with tasks," to clear expectations and to a safe work environment.

For this chapter, NVOAD sets out POC on volunteer management specific to organization-based disaster response in two key ways. First, volunteers have a right to know what an organization does, how it is organized, who leads it, how it manages finances, and how it handles liability. Disaster organizations who affiliate with NVOAD should also inform volunteers about how and where they can volunteer and what skill sets are needed on those sites in order to use volunteer talents "wisely and effectively" (NVOAD, 2011). Member organizations also embrace points of consensus on volunteer housing, from quality to safety matters. Clearly, the umbrella organization function of NVOAD heightens the responsibility levels of affiliated disaster organizations.

In addition to what member organizations should practice, volunteers also share a responsibility to insure that their talents suit the organization's mission and worksites. Part of that responsibility includes being honest about "skill level, experience, and availability" for repairs and reconstruction (NVOAD, 2011, 2013). Volunteers should also be enthusiastic, follow safety requirements, and understand how to do the work they have been asked to perform. They also bear an ethical responsibility to respect the organization and to treat survivors with respect and equity.

It is impressive to watch a NVOAD member organization in action. Mennonite Disaster Service (MDS), for example, works to repair and rebuild the homes of low-income and vulnerable people caught in the aftermath of a natural disaster like tornado or wildfire. They include a paid staff in two locations (the U.S. and Canada) and a specific mission statement. Financial stewardship is overseen by paid staff and a volunteer board. A volunteer coordination system includes a downloadable volunteer guide, project site supervision, and support including housing and food. Being disaster-oriented, member organizations like MDS monitor disasters in advance to determine if the event fits their mission. As a grass-roots based organization, MDS will probably send a volunteer to investigate the situation, identify unmet needs, and recommend opportunities to the bi-national offices. MDS also coordinates with NVOAD as a member organization and participates in affirming the points of consensus. Both managers and volunteers who want a quality experience would want to search for NVOAD member organizations that support the four principles and POCs.

Types of disaster organizations

While the bulk of disaster organizations seem to come from the nonprofit, faith-based or non-governmental organization sectors, in reality a broad array of organizations presents multiple avenues in which volunteers can help. In this section, readers will explore options for both finding useful organizations to recruit into a volunteer plan and for joining as a volunteer.

Faith-based organizations (FBOs)

Faith-based organizations (FBOs) arise out of religious traditions and offer a wide set of post-disaster aid (Phillips & Jenkins, 2010). Many disaster affiliated organizations form from denominations within Christianity, Islam, Buddhism, Hinduism, Taoism, and Judaism. Christianity, for example, includes disaster organizations spanning multiple denominations: Baptist, Lutheran, Catholic, Methodist, Presbyterian, Anabaptist, and more. For example, Buddhists created the Tzu Chi organization that has distributed relief supplies within a context of compassionate care. For the 2018 Camp Fire in California, Tzu Chi offered care to survivors including bedding. Muslims have created ICNA Relief, which spans preparedness, response, and recovery. For recent hurricanes in the U.S., they offered debris removal, cleanup, and food. NECHAMA represents the Jewish response to disasters, with volunteers recently helping Puerto Ricans recover after hurricane Maria through roofing homes and making them more disaster resilient.

A faith basis also offers a useful foundation for recruiting volunteers and generating funding, typically as an extension of how the faith promotes care for others (Box 7.1, a sampling of scriptures from various religious writings). Muslims, for example, practice five pillars of their faith, one of which is called Zakat and requires tithing for the poor. Rabbinic law code also compels Jews to be charitable as a mitzvah or commandment. Christian traditions also address disasters through scripture, by orienting others to be a Good Samaritan who cares for strangers as neighbors. Such universal moral laws, like the Golden Rule, permeate FBOs in general (Kinnier, Kernes, & Dautheribes, 2000) and thus flavor the response of disaster affiliated

Box 7.1 Faith-based philosophies for serving others

Faith compels people to act in certain ways, particularly in caring for others. Universal values seem to spread across all faith traditions, including what some all the Golden Rule: to care for others as you would care for yourself (Kinnier et al., 2000). Doing so requires people to share a cup of water with a stronger, to shelter the downtrodden, and to not bypass the stranger. Religious writings invoke these behaviors, which are passed on through faith-based organizations:

Christianity, Judaism

Love your fellow as yourself. Leviticus 19:18

Taoism

The sage takes care of all men and abandons no one. Tao Te Ching 27

Islam

Pay, Oh Children of Adam, as you would love to be paid, and be just as you would love to have justice! Qur'an 83: 1–6

Buddhism

Hurt not others with that which pains yourself. Udanavarga 5:18

organizations stemming from people's beliefs. Volunteers can nearly always find a FBO focused on disasters and linked to their belief system (for examples, see Box 7.2).

Faith-based organizations tend to identify and respond to unmet needs or underserved groups. Lutheran Social Service of Minnesota in the U.S., for example, created Camp Noah as a way to help children impacted by disasters to process their experiences. Camp Noah partners with an array of agencies and organizations to set up local programs in disaster areas. Their camps last five days and connect peers who have gone through a similar experience. The children learn how to grieve and how to cope, with skills that can last a lifetime. Research shows that learning such coping skills can make people more resilient including with disaster and non-disaster events (Norris, Friedman, & Watson, 2002; Norris, Friedman, Watson, Byrnea, & 2002).

The Uniting Church in Australia demonstrates that the faithful do not always to go to the disaster scene to help. This faith-based denomination raises funds for disaster sites and provides resources like disaster chaplains. When a bombing in Bali, 2002, left Indonesians and Australians dead and injured, the church created materials to help people process grief. They posted memorial services, listed resources, found hymns, and organized prayers. Such emotional and spiritual support is needed, and functions best with people who are properly trained to help. Multiple organizations offer training, which tends to follow NVOAD's Points of Consensus for such care. In general, the purpose of such care is to promote resilience and provide an early aid to help facing trauma. Emotional and spiritual care does not replace psychotherapy which is offered by credentialed professionals (more on this later). Those who provide care emanating from the faith-based community should

Box 7.2 Where can you get training to join an Affiliated Disaster Organization?

The following organizations offer various kinds of training for disaster volunteers. By searching a disaster voluntary organization that interests you, additional training should be easily found. These links were last accessed on July 18, 2019.

American Red Cross

https://www.redcross.org/volunteer/become-a-volunteer.html

Camp Noah

https://www.lssmn.org/campnoah/

CARE

https://www.care.org/emergencies

Center for Disease Control (health care)

https://emergency.cdc.gov/coca/trainingresources.asp

Church World Service Webinars (long term recovery)

https://cwsglobal.org/our-work/emergencies/webinars/

Doctors Without Borders (health care, international humanitarian crises)

(https://www.msf.org/work-msf

Federal Emergency Management Agency (U.S., volunteer management and general emergency management)

https://training.fema.gov/is/

ICNA Relief (preparedness, mitigation, relief work)

https://www.icnarelief.org/disaster-relief/training/

Medical Reserve Corps (health care)

https://mrc.hhs.gov/volunteerfldr/AboutVolunteering,;

New Zealand Ministry of Civil Defence and Emergency Management. (2019). Available at

https://www.civildefence.govt.nz/cdem-sector/capability-development/cdem-training-courses/, last accessed July 29, 2019.

Oiled Wildlife Care Network (see links to partner agencies specific to their mission)

https://owcn.vetmed.ucdavis.edu/readiness/training

Operation Blessing (general volunteers and heavy equipment operators)

https://www.ob.org/disaster-relief/volunteer/

Red Cross/Red Crescent (multilingual, shelter, mass care)

https://www.ifrc.org/en/get-involved/learning-education-training/learning-platform1/

ShelterBox (shelter, logistics)

(https://www.shelterbox.org/operations-update)

Veterinary Medical Assistance Teams (animals)

https://www.avmf.org/whatwedo/veterinary-medical-assistance-teams/

World Vets (animals)

https://worldvets.org/campaigns/disaster-response/

Now that I am trained, how do I volunteer? For examples of places that register volunteers in advance, visit https://mds.mennonite.net/volunteer-registration-form/ or https://www.icnarelief.org/volunteer/ or contact the organization of interest through their website or social media.

For a review of various organizations including how they spend donated funds, visit https://www.charitynavigator.org/.

also secure specialized training to do so. Their work must embrace the highest ethical principles of integrity, respect, and accountability and to respect the diversity found within the communities in which they will volunteer.

Certainly, FBOs also send forth large numbers of mission teams as well – full of unskilled and skilled volunteers to address problems in disasters. Many focus on repairs and rebuilding efforts and bring in weekly crews over time to complete such work. They often interconnect and work collaboratively on projects which enables them to pool resources, time, talent, and funds. Individually, FBOs active in disaster represent the bulk of disaster organizations. Collectively, they make a powerful impact by providing a focused, ordered effort.

Civic organizations and civic interests

Civic clubs can be exceptionally useful vehicles through which to organize and funnel both tangible and intangible relief. Civic clubs tend to organize around similar interests like business professionals, fraternities and sororities, public service, fire and police or even sports and recreational interests (Finch, 2016). Such clubs might be just local or could enjoy a larger reach spanning the region, nation, or planet, like Rotary International efforts to eliminate polio. Local networks, like the kinds generated through community-based ties enjoyed in civic clubs, prompt trust because residents can see the good works in which civic club members participate. Other residents in those communities know who belongs to civic clubs and become more likely to share, volunteer, or donate to those they trust. Canadians have encouraged the incorporation of civic groups and their invaluable networks into emergency management volunteering, a trend that has been increasing through programs embracing corporate philanthropy and paid time off to volunteer (Waldman, Yumagulova, Mackwani, Benson, & Stone, 2017).

Communities worldwide enjoy the benefits of civic organizations like Rotary, Elks, and Lions Clubs among others. Their collective work comes from members who share a set of common beliefs. Rotarians worldwide, for example, emphasize behaviors that value truth, fairness, beneficence, and goodwill. Australian Rotarians support Disaster Aid, which has provided safe drinking water projects and earthquake relief in other countries. Rotarians also do more than projects, such as raising money for immediate relief to Mozambique after cyclone Idai in 2019 (Oxfam, 2019). Lions Clubs focus on eyesight, so it is not surprising that they fund vision clinics or mobile units in order to provide new glasses lost or damaged in a disaster. Lions have also funded clothing and food as well as long term reconstruction.

Civic clubs reflect rather diversified interests, as evidenced after the terror attacks on September 11, 2001 and the 2013 Boston marathon bombing (Finch, 2016). Sports club members visited those who had been harmed, raised money, and participated in marketing to promote resilience. Professional teams became the first to re-establish normal routines after these tragedies by restarting their seasons which brought people together both at the ballparks and beyond. At a post-Katrina conference in Louisiana, I noticed the wait staff from the banquet had disappeared. I happened to find them down the hallway, gathered around a television. They were watching the New Orleans Saints re-take the field in the newly repaired SuperDome, which had become a last-resort shelter from the storm. The Saints went on to win the Super Bowl that year, in a redemptive, joy-filled reclaiming of the city's spirit. Civic clubs and those oriented toward the public good have the potential to pull people forward out of disasters.

To further illustrate civic-mindedness, Team Rubicon emerged after the 2010 Haiti earthquake, arising out of military veterans searching for a meaningful mission. Emphasizing service above self, veteran volunteers embrace tenacity, impartiality, accountability, collaboration, and innovation as core values. Initially organized by U.S. marines, Team Rubicon has since created U.S. and international volunteer opportunities to donate medical skills, logistics expertise, and humanitarian leadership with and for the United States, Canada, Pakistan, Nepal, Chile, Sierra Leone, the Philippines, and Indonesia. Partnerships to create local Team Rubicon units have since evolved in Norway, Australia, the United Kingdom, and Canada.

Community organizations active in disaster (COADs)

Another way in which people and places address disaster risks is through Community Organizations Active in Disaster or COADs. Typically comprised of both public and private agencies and businesses, COADs may work on any aspect of the disaster life cycle. Their goals typically include managing community-based disaster risks and promoting resilience (Maskrey, 2011). Organizations active in a COAD may represent a range of interests, but ultimately try to deliver information and resources to community members. That work might include preparedness information and education on how to prepare a go-kit, to shelter in place, or to evacuate (Ainuddin & Routray, 2012; Chandra et al., 2013). COADs might also become involved in disaster planning, training, and exercises.

That is exactly what the Pike-Ross-Hocking Health Care Coalition has done in the state of Ohio. The effort brought together people representing extended care facilities, home health care agencies, hospitals, emergency management and public health agencies and the Red Cross to plan for an untoward event. In 2018, they trained collectively to address functional needs populations including seniors and people with disabilities. As their training scenario, they worked through issues surrounding an extended power outage in an ice storm. By discussing what they would do and how they would do it, participants surfaced issues to take back to their units from personnel transportation and management to generators that would sustain life. This local COAD represents best practices for cross-community collaboration in advance of a disaster.

That kind of successful effort is not always the case with health care agencies, which tend to push out information through partners rather than build a coalition (Stajura et al., 2012). To engage with a broad set of agencies requires time and commitment, as well as the leadership necessary to make it happen which can be difficult given the historically heavy workload of agencies like those tasked with public health. Leadership requires networking and convincing other partners to come to the table. Those partners must learn to communicate across their different cultures and perspectives to reach a common goal. They must also trust each other, which takes time to develop (Chandra et al., 2013; Nolte & Boenigk, 2011; Stajura et al., 2012). Such work may take place on top of an already full workload, especially for public health agencies. Still, many places lack agencies or sufficient personnel to do the local work of disaster management. Local partners and community entities can be key helpers to stretch limited government resources as an added benefit of voluntarily collaborating and connecting to disaster organizations that train cadres of volunteers (Ainuddin & Routray, 2012).

COADs may be particularly helpful with disadvantaged communities, so that local circumstances, cultures, languages, and situations are fully understood. Linking COADs to true community-based efforts, like neighborhood associations or specific interest groups, can yield even higher benefits (Berke, Cooper, Salvesen, Spurlock, & Rausch, 2011). Not everyone has the means to evacuate, for example, nor do they always use the channel(s) through which information is disseminated. Spurring people to action also relies on trust between locals and agencies. COADs can increase the range of trusted people who connect to authorities and agencies with resources. That was the case with hurricane Katrina, when health care agencies and medical professionals encouraged people to evacuate (GAO, 2006). When communities, governments, and nongovernmental organizations collaborate through a largely volunteer effort, lives can be saved and communities can be restored.

Non-Governmental and International Non-Governmental Organizations (NGO, INGO)

In contrast to a coalition like a COAD, NGO and INGOs may act alone or potentially in concert with each other. What typically distinguishes an NGO is that they operate as a nonprofit focused on humanitarian relief in a disaster situation and rely on volunteers. NGOs typically operate separately from government though, again, they can work with government partners to supplement what locally affected communities need.

For example, Haiti was not the only governmental system to be overwhelmed when disaster struck. The 2011 tsunami in Japan devastated 227 municipalities, killed over 19,000 people and injured over 6000. Hundreds of thousands faced displacement and entire towns had to be rebuilt. The tsunami destroyed the Otsuchi town office and killed 24% of the governmental employees including the mayor (Aoki, 2016). Rikuzentakata City lost 23% of its staff as well. Involving NGOs represented a new approach in Japan, but this catastrophic disaster produced such a need. One nonprofit, the Entrepreneurial Training for Innovative Communities created a "Right Arm Fellows" program to support the loss of governmental and private sector employees (Aoki, 2016).

Because governments cannot always handle massive humanitarian needs after disaster, International Nongovernmental Organizations (INGOs) operate across nations, providing a global reach through their efforts. One INGO, ShelterBox, works in most continents providing resources specific to the disaster, climate, and culture. Recently, ShelterBox provided tents and blankets to Syrians as well as kitchen sets, traps, and sleeping blankets to flood survivors in Ethiopia. In Paraguay, the organization offered flood victims an array of resources including tarps, solar lights, and mosquito nets. Such a focused approach matters when serving areas affected by disaster so that – again – the right items get to the right people in the right places at the right time. Volunteers fuel those efforts as do the mindful and intentional contributions of donors. INGOs enjoy monetary support from individuals, other organizations, and the private sector, which also steps up in other ways.

The private sector

Undoubtedly, civic, faith-based, and other service units benefit from the private sector, particularly in the way of in-kind donations and money. Businesses and corporations of all sizes donate generously during disasters and also fund volunteer organizations through their foundations and philanthropic endeavors (U.S. Chamber of Commerce, 2019). In Canada, one report found that 84% of donating companies gave cash, not including raising funds through employee and customer donations (Ayr, 2018). In addition, 30% donated goods and products. Tim Horton's, a popular Canadian coffee store, raised $800,000 Canadian dollars through donut sales for those affected in the Humboldt ice hockey team bus crash.

The U.S. Chamber of Commerce (2019) reported that companies donated at least $405 million specifically to disaster relief in 2017. Americans affected by hurricanes Maria, Harvey, and Irma found disaster relief supported by hundreds of businesses and corporations. Donations from the private sector also helped those harmed by the shooting in Las Vegas and wildfires in California. Funds also went worldwide, with relief money going to earthquake survivors in Mexico City and other disaster-affected locations.

In-kind donations have poured out too, with help reflecting the specific capabilities of private enterprises (U.S. Chamber of Commerce, 2019). Large companies like UPS (United Parcel Service) flew meals, tarps, and shelter supplies to Puerto Rico after hurricane Maria. Beverage corporations and beer companies changed their distribution systems to can or bottle water. IBM focused its corporate energy on small and medium businesses by providing mobile and cloud disaster response capabilities. An Australian biotechnology company donated tetanus vaccines to the Philippines.

Private sector companies also organize their own volunteer efforts. Home Depot employees volunteered to pack cleaning supply buckets in concert with Convoy of Hope and Hero Box for 2018 disasters in the U.S. A Canadian company, TELUS, sent employee volunteer teams to the Fort McMurray wildfires, created a Text2Donate campaign for their customers, and offered phones and chargers to survivors. Lowe's (2019), another home construction supply store, enables employees to travel to affected communities to help. Although it may be difficult to leave one's place of work, by working within the framework of fundraising, resource acquisition, work-based service and digital support, employees within the private sector can become volunteers. Increasingly, corporate philanthropy includes fundraising frameworks and service opportunities as well. It would be worthwhile for prospective volunteers and emergency/volunteer managers to look into their own places of employment for such resources and opportunities or to inspire their employers to develop programs.

Professional organizations

Professional expertise is always needed for disaster operations, including volunteer efforts. Professionals bring in a carefully cultivated background suitable for specific and sometimes high intensity tasks: medical care, dentistry, optometry, mental health, and technology, for example. Typically, professionals bring a combination of credentials earned from degrees or certifications along with licenses that permit them to offer health care, repair electrical systems, fly planes, provide health care and more. Coupled with experience, they can leverage their knowledge for affected communities (see Box 7.3).

To illustrate, Air Care Alliance involves about sixty voluntary pilot organizations in medical and humanitarian transportation. Many of the associated organizations focus on transporting patients with various medical needs. Some, like Operation Airdrop and the Emergency Voluntary Air Corps, concentrate on disaster relief by moving passengers and donated materials into and out of disaster areas. Angel Flight, normally focused on medical needs patients, supported governmental operations during hurricane Katrina to

Box 7.3 Professionals in disaster service

Professionals interested in providing disaster volunteer help may need to pursue training in addition to their usual credentials. For the Red Cross, Disaster Mental Health team members must hold a license that was produced through master's level or higher work. This applies to psychiatrists, social workers, psychologists, counselors, school counselors and school psychologists. Registered nurses who specialize in and have a certification for psychiatric nursing can volunteer (Rn-BC, PMHNP-BC or PMHCNS-BC). Professionals must also hold a license in the state where they live. UK-MED prepares teams for international service through training pathways. Their volunteers include anaesthesiologists, rehabilitation specialists, pharmacists, spinal specialists and others.

Sources: For details, visit https://www.nasponline.org/resources-and-publications/resources-and-podcasts/school-climate-safety-and-crisis/mental-health-resources/become-a-red-cross-disaster-mental-health-volunteer; https://www.apa.org/practice/programs/dmhi/involved and https://www.uk-med.org/?page_id=262.

reunite children separated from their families (Broughton, Allen, Hannemann, & Petrikin, 2006). Another, Pilots to the Rescue, moves animals out of harm's way by transporting them out of shelters near areas that could flood. Doing so enables those same shelters to take in animals subsequently displaced by the event.

And, imagine the full array of animals in need: small pets, farm livestock, zoo exotics, aquariums, and wildlife. Veterinarians, veterinary students, and technicians volunteer through their professional organizations too. The non-profit group World Vets responded to flooding in Nicaragua in 2017 and to Nepal's earthquake areas in 2015. Their teams of professionals included veterinarians, technicians and trained assistants, provided water, food, medications, animal care, financial support, and capacity building for damaged areas. Australians routinely provide care to slower-moving animals, like koalas, at risk in rapidly-spreading wildfires. In the U.S., local or regional animal rescue teams have developed from local county animal rescue teams) to regional Veterinary Medical Assistance Teams (VMATs) that respond nationally. It is much more challenging to safeguard larger wildlife like elephants or dolphins, which would require specialized environments and handling to help them. Responding to evacuate them or provide medical care would also require experience and expertise in order to avoid further risks to people and animals.

Marine life affected by oil spills have required both lay labor and well-prepared animal experts who will help with cleaning wildlife and providing medical care. The California Department of Fish and Wildlife invites volunteers who are properly trained to participate in spill cleanups and rescues. Their efforts link to the Oiled Wildlife Care Network (OWCN), a COAD that includes forty participating organizations (agencies, nonprofits, and universities) across the state. OWCN includes pre-trained volunteers as well as individuals affiliated with response organizations and even spontaneous volunteers.

Specialized areas in communities may also require expertise. Wildfires or floods, for example, damage environmental resources, destabilize hills and cause landslides, push refuse into estuaries, further endanger threatened species, and destroy plants and food sources. A wide range of seasoned professionals may be needed: botanists, foresters, biologists, sustainability experts, zoologists, park rangers, geologists, hydrologists, ichthyologists, and hazardous materials specialists. The National Alliance of State Animal and Agricultural Emergency Programs in the U.S. has created a series of best practices and standards for a range of animals subject to evacuation, injury, sheltering, and treatment. In addition to the expected expertise, such as what veterinarians would offer, participants must also learn the incident command system, the national incident management system protocols, and the national response framework to fit into medical and related care systems for both fauna and flora. Clearly, they will be ready to integrate into a similarly organized effort even in an unfamiliar area.

Health care may indeed represent the most commonly relied-upon form of professional care in disaster situations, including for humans. Usually focused during response operations, medical care spans physical, dental, and mental health care. Doctors without Borders involves paid staff and volunteers in medical humanitarian assistance to disaster survivors. Opportunities exist to support this kind of effort across a range of abilities: medical, finance, logistics, administrative, human resources, communications, technology and more. Dental teams, sometimes through private offices and at times through professional organizations, offer routine care, which may be of considerable value to people with dental injuries from a disaster or who have lost dentures. Dentistry professionals also offer mobile clinics in disaster sites and support local practices that have been affected by a disaster. One unique aspect includes forensic odontology through which dentists can help to identify the deceased (Prajapati et al., 2018). Organizations like the British Psychology Society (a registered charity) provide information and training to work with those affected by terrorism, military trauma, disaster impacts, and experiences that propel refugees and asylum seekers to find safe havens. Various Red Cross and Red Crescent units involve psychology professionals in direct care and require that they have independent licensing (such as a degree in psychology, psychiatry, social work, or counseling) and related local or state licenses and certifications. As previous disasters have revealed, care must be taken to offer psychological care that is not only professionally-informed but also culturally-appropriate and situated in people's faith systems, which promotes healing (De Silva, 2006; Gillard & Paton, 1999).

As another form of care that promotes well-being, faith traditions offer spiritual and emotional care (Roberts & Ashley, 2008). NVOAD's Points of Consensus on this subject outlines the need for training and credentialing in order to deliver appropriate emotional and spiritual care. Faith-based care also includes more than individual support, such as providing rituals that bring comfort (e.g., communion, meditation, chanting, singing) and aiding people through grieving for families, homes, neighbors, communities, pets, and places with funerals, counseling, and providing comfort care (Roberts & Ashley, 2008). Religious clergy also supply labor to areas that have lost their faith leadership due to injuries, deaths, or displacement and provide respite support so that local clergy can regroup and engage in self-care. Losing a worship site can be traumatic as well, because they serve as collection points for people who need support or want to help (Cain & Barthelemy, 2008).

Not surprisingly then, the built environment also requires expertise. Damaged buildings will need to be assessed by engineers, architects, electricians, plumbers, planners, and building code specialists for general safety and before repairs, reconstruction, and rehabilitation can occur. The built environment also includes ports and airports, roads, railways, and highways, dams and levees, utilities and cyberspace. Do paid professionals exist to do this kind of service? Indeed they do, and some arrive as volunteer teams. It is not unusual after a disaster to see a "swarm" of utility vehicles descending on a stricken area. Those swarms will include paid staff as well as volunteer professionals licensed and certified to conduct utility work. Within those swarms can often be found people whose own communities experienced harm and came to pay it forward. A final aspect of the built environment includes telecommunications and cyberspace. Cellular companies will often provide quick and free access to cell charging and portable cellular towers while repairs are being made. One unique effort, Technologies Sans Frontiéres (TSF), focuses on connectivity and access to technology. In Africa, TSF includes disaster risk management as one of their non-profit initiatives linked to the Kyoto Protocol for disaster risk reduction worldwide. TSF also responded to Nepal and helped to set up emergency satellite communications.

Saving buildings and world heritage treasures also matters to historic and cultural preservation enthusiasts. In addition to disaster preservation efforts by the UNESCO World Heritage Site initiatives, similar efforts exist across a number of countries. Both rapid and slow-moving disasters threaten such sites. Terrorism and war have destroyed many priceless artifacts in recent years, with professionals going to great lengths – including losing their lives – to save priceless manuscripts and art. Other forms of disaster, such as climate change, threaten coastal treasures both in situ and in museums (Wells et al., 2014). Professional associations offer expertise in how to protect and restore damaged treasures. Historic preservationists share a knowledge about the proper temperature for rare and ancient manuscripts and environmental scientists understand the potential for acid rain to damage statuary.

So, many opportunities exist across a wide range of organizations. Given the breadth and depth of such opportunities, how do volunteers fit in? Can anyone help even people without medical degrees or veterinary expertise? Absolutely. The next section walks readers through the possibilities that people present to disaster organizations, emergency managers, and volunteer coordinators. If you are a volunteer reading this book, you should be able to find yourself here as well.

Types of affiliated volunteers

Three main types of volunteers appear besides those who are spontaneous or affiliated with disaster organizations. When managers and organizations handle these recruits effectively, they can produce considerable results. In this section, we look at how to work with unskilled, skilled, and professional volunteers through organized efforts preferably with pre-departure training. Insuring that people are ready to serve not only enhances the results but also provides an extra bonus of helping to retain volunteers (see more on this in Chapter 9).

Unskilled volunteers

Unskilled labor represents the bulk of spontaneous volunteers yet they can make a considerable difference. Mission teams from faith-based organizations are often composed of unskilled but motivated workers who lack prior experience but, with supervision and some professional support, can rebuild entire homes. Pre-departure training can help ready them for disaster sites, such as volunteering for a local agency that serves people and places in need. Disaster volunteers might want to learn on a Habitat for Humanity site, volunteer in a neighborhood clean-up, or do small repairs for seniors through local clubs and agencies. People interested in disaster animal care could volunteer at a local rescue or foster animals waiting for their forever homes. Disaster organizations also provide pre-departure training in person or online, and some do so at the disaster site (see Box 7.2).

One organization of many that provides pre-departure training is the Medical Reserve Corps (MRC), which supports routine and disaster-time public health needs (see Raja, 2019). MRC trains its volunteers individually and collectively, as well as both online and in person. Volunteers for MRC also learn by doing, through participating in drills and exercises

(Gist, Daniel, Grock, & Lin, 2016). As an example, a MRC unit (local, state, or national) might practice setting up a Point of Distribution (POD) for a health care crisis like a pandemic. Or, they might practice opening a point of entry for evacuees arriving from a disaster zone. Though MRC does embrace its health care mission by recruiting professionals from relevant fields, people can also volunteer from outside of the profession. Setting up a medical care unit after a disaster, for example, takes people with office skills, logistics, and case management interests (for the latter, see Kamrujjaman, Rusyidi, Abdoellah, & Nurwati, 2018). MRC not only trains people for health care, they require emergency management related training. Certified MRC volunteers, for example, would know how to use the incident command system (ICS), particularly the key ways that ICS that enables strangers to organize and communicate on a new disaster site and to set up operations quickly and efficiently.

Either before or once on site, a standard best practice with unskilled volunteers is to have them fill out a quick form or online survey indicating prior experience such as painting, cleanup, or cooking or at least a willingness to learn and participate in a team. Project supervisors of disaster organizations (the skilled volunteers with experience) can then assign volunteers into work teams tasked with specific projects. Quite often, volunteer teams go for a week of service with many disaster organizations which allows them to learn that skill and others. Some will return as future volunteers, take on longer-term assignments, and perform skilled labor. A clear benefit, then, of joining an affiliated organization is to be able to increase one's skill set for future disasters. Volunteers often find those same skills useful in their own homes and communities as a side benefit of serving.

Skilled volunteers

Another type of volunteer is the person who comes with a usable skill set and can step into needed work or supervise unskilled helpers. Thinking widely, disaster organizations may tap into a wide range of skill sets: construction, electrical, plumbing, child care, animal support, health care, disability advocates, rehabilitation experts, elder care specialists, technology and computer scientists, and people with language skills.

Disaster organizations would typically want to screen skilled volunteers to insure they hold the proper credentials, certifications, licenses, or experience to help on a site. Typically, skilled labor might also be brought in at a particular time when a rebuilding site is ready for their expertise. Putting in the plumbing, electrical, countertops, insulation, or drywall has to occur at particular times in a (re)construction process. Affiliated, skilled volunteers add value when they can arrive at the right time to keep a project going to bring a family home in a timely manner.

As one example, Operation Blessing uses skilled volunteers to operate heavy equipment. This organization has a crane team capable of departing quickly to volunteer. Their team leaves with 20-ton and 18-ton vehicles to clear debris on roads or move trees from homes. A heavy equipment uses skid steers and loaders to manage smaller tasks that still require skilled labor. They also bring in licensed trades like plumbing and electricians. Disaster organizations can do this kind of focused, intentional work because they see the bigger picture in disaster. Repairing or rebuilding a home takes a considerable and coordinated effort. While unskilled teams can pick up debris, move out furniture, and tear out insulation, skilled volunteers will need to arrive at the right time to keep the project going. Disaster organizations manage the paperwork, building permits, and work crews to make that happen. When the home is ready for plumbing or electrical work, their volunteer skilled labor will arrive on time to keep the project moving and return a family to their home.

Professional organizational volunteering

Skilled volunteers may very well include health care professionals who join units like the Medical Reserve Corps. In addition, they may operate out of professional associations or organizations that task them with disaster-specific work. Let's consider some examples of what kind of professional volunteering might be of value in a disaster, and link it to suitable, affiliated organizations (see Box 7.2):

Consider, as an example, what Haiti needed after the 2010 earthquake:

- Public sector officials, due to significant loss of life in the Haitian government and United Nations compound.
- First responders, because police and fire units suffered numerous casualties requiring additional outside support.
- Experts capable of repairing and re-opening the harbor and airport, along with harbormasters and air traffic controllers able to coordinate massive numbers of relief planes.
- Pilots for large ships and airplanes to bring in cargo and relief teams for years.
- Airplane mechanics and ship maintenance teams.
- Technology teams to restore communications from cellular service for information, email, and social media, which was used to convey emergency messages and from which digital volunteers obtained actionable information for on the ground teams.

- Digital volunteers with geographic and computer skills to map out emergencies and provide information to rescuers.
- Highly experienced search and rescue teams, with canine units, capable of difficult extrications in a highly unstable environment.
- Surgeons to address traumatic injuries and amputations. The U.S.S. Comfort, a hospital ship, provided operating rooms and intensive care units. Health care teams also needed x-ray facilities and technicians along with pharmacists and pharmaceutical supplies.
- Mortality management teams for the estimated 200,000 + who died.
- Security specialists to address incidents of violence against women and girls in the massive tent cities (Duramy, 2012).
- Translators capable of communicating across multiple languages: French, Creole, English, Spanish and languages used by volunteer teams from dozens of countries.
- Rehabilitation experts able to support newly-disabled earthquake victims (Danquah et al., 2014).
- Psychologists with experience in providing trauma care.

Professional involvement in disasters will require various levels of education, certifications, credentials, licenses, and training (see Box 7.3). Someone who provides support to a town's administration may need both practical experience and a public administration degree, a background in finance, or perhaps an ability to write grants and contracts. A psychologist will need training not only involving credentials, but knowledge about how to treat patients within the context of a traumatic experience. Nurses, physicians, and veterinarians will all need education, experience, and licenses. In the UK, a medical organization called UK-MED prepares such people through training pathways that include working in austere and challenging conditions. Disaster organizations prepare their volunteers to serve in the environments generated by tornadoes, earthquakes, and floods. And that is the true value of affiliating with a disaster organization – to be ready for what volunteers will face so they can do the very best they can for those who have survived.

Conclusion

Disaster-stricken communities need not only volunteers, but a concerted and focused effort that enables them to return to normal. By joining an experienced disaster organization, volunteers can help leverage their collective energies to produce the best results. Those organizations, if they follow best practices and collaborate appropriately, can transform the situations of individuals, neighborhoods, and communities. Together, their efforts can take the broken pieces of the human spirit and the places where they live, and

Box 7.4 Building connections with affiliated volunteers

How can emergency managers and volunteer coordinators build relationships with disaster experienced organizations? Here's some steps that could work:

- Join the VOAD movement, perhaps by attending local or state-level meetings or by organizing an area VOAD. A similar effort can occur through coalitions, such as a health care coalition or set of professional organizations.
- Attend conferences of volunteer organizations and invite them to attend emergency management conferences. Get to know each other: personnel, mission, capabilities, procedures, and experience.
- Have coffee with each other, because while formal inter-organizational relationships provide structure and clarity, informal relationships matter the most when counting on someone for disaster relief help.
- Cross train with each other's organization. An emergency manager would benefit from going through domestic violence training as would any shelter manager. Likewise, those with experience in domestic violence in disasters would benefit from emergency management training if only to learn the key words and approaches used.
- Generate formal MOUs outlining each other's capabilities and commitments.
- Develop plans for integrating voluntary organizations and for managing spontaneous volunteers. Disaster organizations should be involved in those conversation to craft a plan suitable to everyone's expertise, experience, and availability.
- Training, drills, and exercises always help test a plan and enable people to get to know each other. Involve voluntary organizations in such events or ask to be invited to the table.

restore well-being. Volunteers, emergency managers, and volunteer coordinators can find a wide range of organizations interested in helping in disasters from true community-based organizations to those formed through corporate philanthropy. Because governments cannot address all needs or because they may lack the resources to do so, disaster affiliated organizations and their volunteers will always be needed. Emergency managers and volunteer coordinators would be wise to become familiar with these organizations and build strong, meaningful connections (see Box 7.4)

References

Ainuddin, S., & Routray, J. (2012). Institutional framework, key stakeholders and community preparedness for earthquake induced disaster management in Balochistan. *Disaster Prevention and Management, 21*(1), 22–36.

Aoki, N. (2016). Adaptive governance for resilience in the wake of the 2011 Great East Japan earthquake and tsunami. *Habitat International, 52*, 20–25.

Ayr, S. (2018). *Corporate giving*. Toronto, Canada: Imagine Canada.

Berke, P., Cooper, J., Salvesen, D., Spurlock, D., & Rausch, C. (2011). Building capacity for disaster resiliency in six disadvantaged communities. *Sustainability, 3*, 1–20.

Cain, D., & Barthelemy, J. (2008). Tangible and spiritual relief after the storm: The religious community responds to Katrina. *Journal of Social Service Research, 34*(3), 29–42.

Broughton, D., Allen, E., Hannemann, R., & Petrikin, J. (2006). Reuniting fractured families after a disaster: The role of the National Center for Missing & Exploited Children. *Pediatrics, 117* (5), S442–S445.

Chandra, A., Williams, M., Plough, A., Stayton, A., Wells, K., Horta, M., & Tang, J. (2013). Getting actionable about community resilience: The Los Angeles County community disaster resilience project. *American Journal of Public Health, 103*(7), 1181–1189.

Danquah, L., et al. (2014). Disability in post-earthquake Haiti: Prevalence and inequality in access to services. *Disability and Rehabilitation, 37*(2), 1082–1089.

De Silva, P. (2006). The tsunami and its aftermath in Sri Lanka: Explorations of a Buddhist perspective. *International Review of Psychiatry, 18*(3), 281–297.

Duramy, B. (2012). Women in the aftermath of the 2010 Haiti earthquake. *Emory International Law Review*, 1193.

Dynes, R. (1974). *Organized behavior in disaster*. Lexington, MA: Health Lexington Books.

Finch, B. (2016). Boston sport organizations and community disaster recovery. *Disaster Prevention and Management, 25*(1), 91–103.

Gillard, M., & Paton, D. (1999). Disaster stress following a hurricane: The role of religious differences in the Fijian Islands. *Australasian Journal of Disaster, 2*, 2–9.

Gist, R., Daniel, P., Grock, A., & Lin, C. (2016). Use of Medical Reserve Corps volunteers in a hospital-based disaster exercise. *Prehospital and Disaster Medicine, 31*(3), 259–262.

Government Accountability Office. (2006). *Disaster preparedness: Preliminary observations on the evacuation of vulnerable populations due to hurricanes and other disasters*. Washington DC: GAO, GAO-06-790T.

Kamrujjaman, M., Rusyidi, B., Abdoellah, O., & Nurwati, N. (2018). The roles of social workers during flood disaster management in Dayeuhkolot District Bandung Indonesia. *Journal of Social Work Education and Practice, 3*(3), 31–45.

Kinnier, R., Kernes, J., & Dautheribes, T. (2000). A short list of universal moral values. *Counseling and Values, 45*, 4–16.

Lowe's. (2019). Jumpstarting recovery after disaster. Available from https://newsroom.lowes.com/inside-lowes/jumpstarting-recovery-devastation/, Accessed 15.07.19.

Lund, F., & Vaux, T. (2009). Work-focused responses to disasters: India's Self Employed Women's Association. In E. Enarson, & P. Dhar Chakrabarti (Eds.), *Women, gender & disaster: Global issues and initiatives* (pp. 212–223). New Delhi, India: Sage.

Maskrey, A. (2011). Revisiting community-based disaster risk management. *Environmental Hazards, 10*, 42–52.

National Voluntary Organizations Active in Disaster (NVOAD). (2011). Points of consensus: Volunteer management. Available from www.nvoad.org, Accessed 10.07.19.

National Voluntary Organizations Active in Disaster (NVOAD). (2013). Points of consensus: Cleanup, repair and rebuild. Available from www.nvoad.org, Accessed 10.07.19.

Nolte, I., & Boenigk, S. (2011). Public-nonprofit partnership performance in a disaster context: The case of Haiti. *Public Administration, 89*(4), 1385–1402.

Norris, F., Friedman, M., & Watson, P. (2002). 60,000 disaster victims speak: Part II. Summary and implications of the disaster mental health research. *Psychiatry (Edgmont), 65*(3), 240–260.

Norris, F., Friedman, M., Watson, P., Byrne, C., Diaz, E., & Kaniasty, K. (2002). 60,000 disaster victims speak: Part I. An empirical review of the empirical literature, 1981–2001. *Psychiatry (Edgmont), 65*(3), 207–239.

Oxfam. (2019). Cyclone Idai in Malawi, Mozambique and Zimbabwe. Available from https://www.oxfam.org/en/emergencies/cyclone-idai-malawi-mozambique-and-zimbabwe, Accessed 11.07.19.

Phillips, B., & Jenkins, P. (2010). The roles of faith-based organizations after hurricane Katrina. In R. Kilmer, V. Gil-Rivas, R. Tedeschi, & L. Calhoun (Eds.), *Helping families and communities recover from disasters: Lessons learned from hurricane Katrina and its aftermath* (pp. 215–238). Washington DC: American Psychological Association.

Phillips, B., Mwarumba, N., & Wagner, D. (2012). *The role of the trained volunteer. Cole and Nancy Connell.* Wiley and Sons.

Prajapati, G., Sarode, S., Sarode, G., Shelke, P., Awan, K., & Patil, S. (2018). Role of forensic odontology in the identification of victims of major mass disasters across the world: A systematic review. *PLoS One, 13*, 6, https://doi.org/10.1371%2Fjournal.pone.0199791.

Raja, K. (2019). The Medical Reserve Corps: Volunteers augmenting local public health preparedness and response. *Journal of Public Health Management and Practice, 25*(1), 95–97.

Roberts, S., & Ashley, W. (2008). *Disaster spiritual care: Practical clergy responses to community, regional and national tragedy.* Woodstock, VT: Sky Lights Path Publishing.

Stajura, M., Glik, D., Eisenman, D., Prelip, M., Martel, A., & Sammartinova, J. (2012). Perspectives of community- and faith-based organizations about partnering with local health departments for disasters. *Environmental Research and Public Health, 9*, 2293–2311.

U.S. Chamber of Commerce. (2019). Corporate giving through the lens of disaster response. Available from https://www.uschamberfoundation.org/blog/post/corporate-giving-through-lens-disaster-response, Accessed July15, 19.

Waldman, S., Yumagulova, L., Mackwani, Z., Benson, C., & Stone, J. (2017). Canadian citizens volunteering in disasters: From emergence to networked governance. *Journal of Contingencies and Crisis Management, 26*, 394–402.

Wells, J., Kansa, E., Kansa, S., Yerka, S., Anderson, D., Bissett, T., ... DeMuth, R. (2014). Web-based discovery and integration of archaeological historic properties inventory data: The Digital Index of North American Archaeology (DINAA). *Literary and Linguistic Computing, 29* (3), 349–360.

Further reading

Habitat for Humanity. (2018). Hurricane Maria recovery. Available from https://www.habitat.org/impact/our-work/disaster-response/hurricanes/hurricane-maria, Accessed 11.09.19.

National Voluntary Organizations Active in Disaster (NVOAD). (2015). Points of consensus: Quick reference guide to the National VOAD disaster and spiritual care guidelines. Available from www.nvoad.org, Accessed 16.07. 19.

CHAPTER 8

Managing international volunteers

Introduction

Reports filtered out of Myanmar slowly after Cyclone Nargis in 2008 when a record storm surge pushed inland across the low-lying Irrawaddy Delta. Refusing aid, the ruling military junta endured worldwide condemnation while people suffered horribly. Millions lost their homes and livelihoods, food prices rose, and around 138,000 people died. The junta first accepted aid from the Bangladesh and Indian militaries rather than from western nations. Other countries then funneled in generators, tents, medical support, food, cash and related aid. With the media denied access or threatened with imprisonment, rumors circulated that donations failed to reach those in greatest need and had been taken by the junta. The United Nations, leaders of many countries, and people worldwide expressed outrage. Some charged that the military junta had a responsibility to protect its citizens and that outside force might have been justified as a life-saving, humanitarian intervention (Barber, 2009; Stover & Vinck, 2008). Clearly, volunteering internationally can prove frustrating and delivering help appropriately can be complicated. This chapter outlines some of the challenges of international disaster volunteering and offers recommendations for preparing and managing international volunteers effectively with a general rule of *do no harm*.

The context of international disaster volunteerism

Context, defined generally as *time, place,* and *circumstances,* influences the ways in which disasters occur and can be managed. Consider, for example, the complexity of place such as an island location, a mountainous area, or a coastal zone. Gaining access would represent the first barrier to address, which can be influenced by the circumstances of particular places. Governments may deny entry out of logistical access problems as well as political or military concerns. In 2004, a longstanding conflict between the Sri Lankan government and the pro-independence Tamil Tigers came at a

Disaster Volunteers. DOI: https://doi.org/10.1016/B978-0-12-813846-5.00008-3

difficult time requiring a temporary ceasefire in order to deliver tsunami aid (Hull, 2009). Thus, just getting there can be insurmountable or dangerous.

Indeed, security issues can affect volunteering in other ways. Due to homeland security policies, Canadian volunteers with Mennonite Disaster Service needed formal paperwork to bring reconstruction tools to the U.S. Gulf Coast after hurricane Katrina. Solar Under the Sun (SUTS), which provides alternative energy sources for water treatment plants, has volunteers carry in separated parts to be assembled later in-country. Weight maximums and U.S. Transportation Security Agency rules limit how they must transport their resources, with required documentation for what could be construed as bomb-making materials. As another example of entry challenges, the U.S. declined offers of international, professional water rescue teams after hurricanes Harvey and Maria in 2017, with the result being an emergent "Cajun Navy" of local, untrained volunteers arriving. Sadly, two such volunteers died when their boat passed over a live electrical line.

Transportation challenges frequently block volunteers into affected places, such as when earthquakes destroy ingress via waterways, airports, and roadways. In Haiti, airplanes had to turn around and land in other countries while militaries made the runway viable. Vehicle convoys then came inland from the bordering Dominican Republic, clogging roadways into affected areas. Reaching the tsunami-hit area of Naggapattinam, India, took a full day by road given limited waterborne access due to port and bridge damage. Helping those harmed by the Japanese tsunami presented even more significant risks, with unknown impacts of radiation from the related nuclear plant problems.

Other circumstances, often unanticipated, also beleaguer those trying to help. Gendered contexts of disasters require attention to often-hidden issues. An earthquake severely damaged Pakistan in 2005 and relief agencies rushed to help. Gendered behaviors meant that men secured food while women, many of them disaster widows, could not obtain sustenance for themselves and their children. Medical care volunteers had to create gender-specific programs to respect norms of modesty and interaction in order to treat women and girls (Sayeed, 2009). The contexts of the 2004 Sri Lanka tsunami as well as the 2010 Haiti and 2015 Nepal earthquakes meant that aid groups trying to deliver basic supplies also had to protect survivors and volunteers from human trafficking and sexual assault (Fisher, 2010). Reconstruction efforts in India after the tsunami disrupted women's roles in rebuilding which cost them income and influence. Such gendered contexts represent embedded social problems known to locals and experienced disaster responders. Outside agencies and organizations must understand such challenging contexts to be effective in an unfamiliar environment. From a social science perspective, a good starting point is to understand the culture in which people want to volunteer.

How culture influences disaster volunteerism

Clearly, international disaster volunteering can challenge even the most experienced relief organizations and political leaders given the context-based challenges mentioned above. Another important part of international volunteering stems from culture, defined as a design for living, which influences what we do, think, say, and feel. Culture underlies how we interact with others, what foods we eat, and the kind of music we enjoy. Culture also informs whether we shake hands or bump shoulders, engage in eye contact or avoid doing so, and use personal space between people. With disasters, culture influences how we make sense of a tragedy, mourn, and move on. People rely on their culture as a way to provide comfort and support and as a framework for organizing relief and recovery efforts. International disaster volunteering thus requires a heightened understanding of the cultures in which people want to help. In this section, we look at various aspects of culture that an individual or team of volunteers must grasp to be effective. A subsequent section then looks at ways in which outsiders have approached other cultures, along a continuum from harmful (ethnocentrism) to helpful (cultural relativism, cultural competence, and cultural humility).

Culture relies on four main elements that emanate from intangible, implicit, and shared understandings: values, norms, language, and symbols. To start, people rely on cultural values to distinguish what they believe to be desirable or not desirable. As an illustration, outsiders often wonder why people stay in locations where they lost their homes and livelihoods, sometimes in an area of repetitive risks. Yet those locations represent a desirable place, where people have formed meaningful attachments to the land, water, livelihoods, places, and people. To lose those locations in a flood or wildfire may mean a loss of one's economic opportunities or the social networks that have enabled them to survive and to thrive. Thus, it is not unusual to hear people say "we will rebuild" even though hurricanes and earthquakes will happen there again. Survivors will refuse to relocate despite known risks, because they value where they live, their abilities to earn a living in the area, and the people with whom they share their lives (Handmer & Nalau, 2013; Phillips, Stukes, & Jenkins, 2012).

People also follow cultural norms, defined as behavioral guidelines, for interaction. Two kinds of norms exist, folkways and mores (pronounced more-ays). Folkways represent less formal ways of interacting, such as how we shake hands or make eye contact (Sumner, 1902). Violating a folkway might be seen as odd or inappropriate, but usually results in minor sanctions such as avoidance of the violator or social distancing. In some cultures, it is appropriate to cover one's head out of commitment to a valued faith tradition. Sanctions for not doing so can range from looks or comments to, in some locations, the potential for significant harm. Similarly, some cultures use the right hand since the unclean left hand is used for washing. Using both hands to eat might horrify local people who would find the behavior disgusting or opt not to eat with such offenders. In contrast, mores outline norms that should never be violated. The examples of human trafficking given earlier represent such severe violations from which people should be valued and protected.

Language also matters to people (Santos-Hernández & Morrow, 2013). Within any given country, one language may be prominent (such as Spanish or French) but in some countries, multiple languages and/or regional dialects may exist. Even when we know a language, such as Spanish, we may not know the local expressions that flavor how people communicate and represent their culture. Costa Ricans speak Spanish, for example, but a common expression is "no te preocupes" which means "relax" or "chill." The intent is to encourage people to be "tranquilo" or tranquil, a core part of the Costa Rican personality. The high stress environment of a post-disaster recovery might challenge tranquility, but working on a return to this culturally-influenced behavior - as expressed linguistically – might also prove therapeutic. In Haiti, misunderstanding erupted over nutritional bars offered by the United Nations. The date of creation stamped on the bars confused survivors who threw away what seemed to be expired food. Language, and related communication skills, matters.

Language also includes non-verbal behaviors. We might wonder if someone looking off into the distance is ignoring us or is overcome by the disaster – until we understand that doing so represents the locally appropriate way to interact. Similarly, knowing how to shake hands per the local norms may also go a long way to making good connections with people we are trying to help. Should you use a firm handshake or not? Or, is it more appropriate to bump shoulders as a greeting? It is appropriate for women and men to touch through a handshake or shoulder bump? Knowing such culturally-influenced behaviors can put people at ease and serve as a way to promote communication, interactions, projects, and helpfulness.

Symbols form one of the more challenging aspects of culture to understand. Symbols, something small that stands for something larger, can vary across cultures. People may refuse to relocate because staying in place symbolizes a historic connection to the land, such as is found within Native American or First Nation communities. Communities may also use symbols to convey

PHOTO 8.1
Sand mandala at a rebuilt home, after the 2004 tsunami, Naggapattinam, India.

stages of recovery. It is not unusual for house blessings to be given when a family can finally return home. Often supported by faith traditions, the blessings typically involve rituals meaningful to people including prayers, incantations, burning of incense, and singing. A symbol of faith may be placed in the home such as a cross (Christianity) or mandala (Hinduism or Buddhism) along with a Bible or Quran. Often, house blessings reflect the core of people's culture as a meaningful and symbolic transition back to normalcy (see Photo 8.1). Working within their cultural understandings helps survivors psychologically and socially.

Failure to understand another culture's design for living can result in an *ethnocentric* response: *why are those people living like that? I would never eat that kind of food! That's not how we do it at home*. Ethnocentric responses to other cultures reflect poorly on volunteer teams and fail to deliver the kind of help that people truly need, embedded in and reflective of culturally familiar and comforting environments. People live "like that" because it is home, where they have imbued places with meaning. What our eyes "see" and our minds interpret may mislead us. I remember listening to a project director talk about the trash in a bayou community, which a recently-arrived mission team had been criticizing. The outsiders assumed that the locals "liked it that way" or did not care about the environment. He gently explained that the bayou had trash in it for multiple reasons, including outsiders who boated through while carelessly littering. Local government had also failed to provide sufficient means for people to dispose of their trash. In contrast, the locals had worked tirelessly as environmental stewards of lands historically owned by Native Americans. The director said, "now that I have been working here and getting to know this community, I don't see the trash anymore – now, I see the people." His insights redirected the mission team's attention to what really mattered – those they had come to serve.

In contrast, a *cultural relativism* approach encourages outsiders to see behaviors and practices relative to local ways of doing something. How a meeting is arranged, for example, varies from culture to culture. Do you set up rows of chairs with a speaker at the front? Or, do you create a "talking circle" where people share space and power? After the Exxon Valdez oil spill near Alaska, talking circles offered a means through which people could come together and feel symbolically safe inside (Picou, 2000). By relying on a traditional and familiar meeting format, participants processed experiences and feelings associated with significant losses to their physical and social environment. They then went on to not only deal with the oil spill but to help their community beyond the disaster. Cultural relativism, understanding people within the context of their way of life, requires that volunteers become culturally competent and practice cultural humility.

Gaining cultural competence on a journey to becoming culturally fluent, essentially a lifelong commitment, takes time and effort. The process of becoming culturally competent requires that volunteers engage in *praxis* by actively and dynamically reflecting on their surroundings and actions (Welty & Bolton, 2017). Reflection might involve journaling, group discussion, and mindfulness. Teams might also provide regular debriefing and discussion around what they are seeing and experiencing. They should think through their own behaviors, which can seem normal to outsiders but offensive to locals. Some groups invite locals to visit with them before and during their volunteer activities as a way to understand and exchange insights on the unfolding efforts. It is not unusual to use food as a way to build connections, forging ties over shared, communal meals with culturally appropriate cuisine unique to the locale.

Practicing cultural humility requires outsiders to put aside their own assumptions and ask about local ways of doing things. We are often concerned about the painful impacts of disasters on survivors. Psychosocial support needs to be particularly sensitive and culturally situated. Consulting and involving local mental health professionals for the appropriate approaches and techniques, and to determine the extent of actual need, is key (Silove & Zwi, 2005). Think about the range of ways in which people need to mourn as a way of moving through grief stages. After the 2004 tsunami, faith influenced many of those mourning rituals including

Muslim, Christian, Hindu, and Buddhist traditions along a five-kilometer stretch of India's affected coastline (Phillips, Neal, Wikle, Subanthore, & Hyrapiet, 2008; Phillips & Thompson, 2013). Familiarity with faith traditions can help promote healing. In contrast, efforts that imported westernized dance and song in a Sri Lankan community affected by the tsunami generated outrage, the opposite of what should have happened (Wickramage, 2006).

Medical teams that volunteered in Haiti brought back useful lessons in terms of cultural understanding. Among their recommendations, working with locals within their practices and contexts proved crucial to prevent friction and foster "mutual respect" (Benjamin, Bassily-Marcus, Babu, Silver, and Marin, 2011, p. 314). Working in a massive relief encampment, for example, required teams to "enter this work with humility and grace" as "guests in another country" (Cunningham & Sesay, 2017, p. 480). By working with people to determine and meet their needs, voluntary organizations can increase their effectiveness and leave a damaged place in far better shape. Local people know their climate and culture best, including the kinds of resources available after external partners leave.

Partnering within a cultural context characterizes a best practice known as *accompaniment*, an approach long-used by experienced humanitarian organizations (Cuny, 1983; Jasparro & Taylor, 2011; Welty & Bolton, 2017). The idea of accompaniment is based on "companionship, active listening, and solidarity" that relies on a "local viewpoint, knowledge and skill base" (Hampson, Crea, Calvo, & Álvarez, 2014). Those who accompany build trust and become advocates. After Haiti, a group called MADRE sent Creole-speaking volunteers to Haiti and coordinated accompaniment with advocacy efforts involving local and external partners (Welty & Bolton, 2017).

Practical realities and best practices

Given the wide array of challenges with international volunteering, it makes sense to engage in a significant amount of pre-departure planning. The starting point may be determining if and when a team should deploy. Too often, well-meaning teams enter too quickly. With so much attention on immediate relief efforts, teams should consider waiting. Recoveries can linger for years or decades, thus in many cases, work will still exist outside of media attention and public concern. Before departure, a number of steps should be taken into consideration including how to make entry into a community, securing necessary paperwork and approvals, insuring health and safety, finding meaningful work, fitting in, and establishing partnerships.

Making entry

The first rule of entrée is to be sure you have been invited, have a clearly-defined mission, and can work with local organizations. Increasingly, disaster-savvy leaders and organizations are requiring international teams to register, train, and credential their volunteers (Merchant, Leigh, & Lurie, 2010). Costa Rica, for example, is among the first to credential emergency medical technicians (EMT) for international deployment using standards established by the Pan American Health Organization (PAHO) and the World Health Organization (WHO). In addition to medical expertise, certified EMTs must be self-sufficient for a minimum of two weeks, arrive self-contained, and not unduly impact locals. In an international disaster context, volunteer teams should not assume that housing, food, hydration, transportation, power, communications, or language support will be available.

Further, while arriving self-contained and ready to step in is important, teams should consider the consequences of how they do this. Haiti suffered terribly from the 2010 earthquake, including major losses to the health care sector (Farmer, 2011). A nursing school collapsed, hospitals fell into ruins, and medical personnel died. A number of well-organized medical teams arrived quickly and set up field operations, surgical theaters, and clinics. While much of the aid mattered and enabled survival, the efforts also displaced local health care workers from their jobs (Jobe, 2011). Eventually, medical mission teams left Haiti with survivors still in need of medical care, including people newly disabled by amputations but without follow-up care, physical therapy, or assistive devices. Volunteer teams should thus consider their impact and think through when to arrive and how need may linger into an extended recovery period. Immediately entering an affected area should be undertaken by experienced teams with an eye toward the appropriate after care teams that may be needed. As noted in previous chapters, help tends to be focused on the response time period when the extended recovery requires support as well.

Another aspect to consider is financial. Traveling into another country can be costly for volunteers and organizations. Imagine instead, how funds spent on travel by a single volunteer might be leveraged by a local NGO. Three thousand dollars spent on a single plane ticket would "pay the average wages of two Haitian construction workers for an entire year" (Welty & Bolton, 2017, p. 116). The funds could also support businesses struggling to regroup, for transporting local workers to job sites, or reuniting families separated by the event. Going in quickly can also cost more money. Since the recovery time period is often under-volunteered, it may be more financially viable to wait while a team finds local partners and plans well-thought out efforts. Waiting may also prove beneficial in terms of risk reduction to volunteers.

Health and safety concerns

International disaster volunteers should check with health care professionals before traveling into disaster areas. Clearly, advance immunizations should be undertaken by arriving volunteers in order to reduce the risks of communicable diseases. Given that immunizations take time to become effective, volunteers need to have them in advance as part of pre-deployment protocols. For example, the U.S. Center for Disease Control requires immunizations for disaster responders including tetanus and hepatitis B. Travel to specific countries might result in additional vaccines. Travel to Haiti should include immunizations against hepatitis A, malaria, and typhoid with additional consideration for areas or jobs (e.g., health care) that might be at risk for cholera, rabies, typhoid, and yellow fever.

Equally important, travelers should not bring contagious diseases or illnesses into an area. After the 2011 Great East Japan earthquake, a disaster volunteer with tuberculosis helped evacuees. From this "index patient," epidemiological investigations revealed that 72 people tested positive for latent tuberculosis infection including 26 earthquake victims (Kanamori et al., 2013). A post-earthquake cholera epidemic in Haiti originated in a Nepalese peacekeeper team's waste disposal area. Seven thousand people died (Frerichs, Keim, Barrais, & Piarroux, 2012). Volunteers should be monitored for illnesses they might bring into a disaster zone as well as those contracted during their time abroad. Even something as simple as daily temperature checks and assessments can make a difference, as recommended after the Japan earthquake and tsunami and to monitor post-disaster conditions (Takahashi, Kodama, & Kanda, 2013). After hurricanes in 2017, PAHO and WHO (2017) launched early efforts to detect diseases that might be spread by water, insects, or rodents.

In terms of health concerns beyond illness, studies remain unclear. Injuries to volunteers go largely unreported in disasters, though it is clear that people do sustain injuries (Miller & Garrett, 2009). A ten year study by the U.S. Center for Disease Control & Prevention (2005) found that some volunteer work carried a higher risk of death, such as firefighting or driving vehicles. Training for volunteer work should certainly include identifying risks and designing safety training. Higher risk volunteering should be of particular concern (Center for Disease Control & Prevention, 2005). Daily updates should be provided so that people remain aware of their surroundings and the related risks. Protective equipment should be provided with supervisory oversight to insure it is donned and used correctly.

In addition, volunteers need to be ready physically for the environment in which they will work (Jobe, 2011; Merchant et al., 2010). They should not travel if they have health conditions that could worsen in a disaster zone or an inhospitable climate. Those dependent on refrigeration for medications or power for durable medical equipment may need to wait or to find alternative ways to volunteer. Fundraising represents one way to help local agencies affected by the disaster both in the immediate response time and into the recovery period. Technologies and social media can also connect volunteers to an area in need including setting up long-term projects, providing expertise for reconstruction, or comforting survivors through spiritual or psychological care.

Finding meaningful work

How does an organization find meaningful work for their volunteers? Here, the presence of experienced organizations and affiliated volunteers truly makes a difference. Organizational missions drive what they do, so looking for the right fit is key. Doctors without Borders, for example, declined funds for the Japanese tsunami because the health care system within the affected nation was already sufficiently robust. They negotiated carefully before deploying personnel to Sri Lanka and Myanmar in the midst of intense conflicts, and reduced risks to their volunteers. Such organizations necessarily spend time conducting needs assessments to discern the fit between what they can offer and what is truly needed. They do not make assumptions or deliver what they think the local country needs. Nor do they impose their own approach on a local community. As discussed earlier, Pakistan's earthquake revealed that a one-size-fits-all approach would prove unsuccessful and that aid organizations must consider, understand, and work around a range of gendered contexts to be successful.

Such work stands in contrast to the "donor-driven" approach, where donors and outside organizations push money into an affected area, assuming that what they have to deliver will be useful. Funders may pressure organizations, both internal and external, into producing something for the money. Yet, owner-driven approaches are more likely to be accepted and embraced (Davidson, Johnson, Lizarralde, Dikmen, & Sliwinski, 2006). People are likely to reject well-meaning help if it does not fit with their needs, context, and environment.

Finding meaningful work for donors, organizations, and volunteers thus requires working with locals and within the local context. Doing so starts from asking "what would be helpful to you?" and listening to what people say. Think of this as a therapeutic strategy, when asking people who have been harmed results in their ability to make empowering decisions that propel them forward into their own recovery.

In the U.S., Long Term Recovery Committees/Groups (LTRGs) operate locally where experienced disaster organizations work with community leaders to surface local needs (see previous chapters). In more than one community, I have heard case managers and recovery leaders talk knowledgeably about the survivors – their neighbors – and who needs help the most. They have also identified clients who would otherwise have gone un-aided and advocated for outside organizations unfamiliar to wary locals. Thus, an effective strategy of outside organizations is listening carefully to exactly what locals need. Those listening skills are deeply appreciated by the LTRG and officials tasked with being good stewards of reconstruction funds and public well-being.

Fitting in

"Those are *our* Mennonites" the Catholic priest said to me during an interview. The volunteers the priest mentioned had come from well outside of the affected town and many had traveled thousands of miles to help from another country. They did not know each other before the disaster, but they felt connected now. As part of their approach, the outsiders had attended services in area churches and worshiped in ways different from their own. They had participated in local events, come to meals provided by local churches, and invited people to their own encampment for food, storytelling, and cultural exchanges.

Finding meaningful work and fitting in flows from several strategies. Making a low impact, self-sufficient entry matters to locals burdened with disaster and struggling to secure their own resources. But such an entry is only an initial step in the process of serving effectively in another cultural context. The idea of service, of being there for others, should permeate the spirit of entry and interaction. By putting aside our own agendas and judgments, we honor those we seek to help and make them feel better by allowing them to be themselves. By embracing who they are and understanding the situation from their perspective, we form a basis to build meaningful and effective partnerships. Part of this means training volunteers to want to learn about the local context and culture and to appreciate what they are experiencing. Just as we want to share our food and customs with international guests, so do the people we want to serve. In Louisiana after Katrina, locals said how much they loved the volunteers eating gumbo, going out on their shrimping boats, and watching ships go down the Mississippi River. One said that it made them appreciate their own community even more, through being able to see it with a fresh set of eyes beyond the lingering storm debris.

Establishing partnerships

Imagine that you are one of the people affected by a disaster. The local agency that you lead needs to be involved but the task ahead is daunting. Then, dozens of aid organizations begin to arrive. Who are they? What do they do? Will they be respectful of your role in the community and the connections you have with neighbors and professionals? Do they speak the same language as you and your employees? Will they help your people? How long will they stay? Can they really deliver on what they offer?

Indeed, the inundation of aid can be overwhelming and dynamic. For Haiti, 396 international agencies registered within the United Nations health cluster system alone. Two weeks after the earthquake, 76 such agencies had already left the country (Goyet, Sarmiento, & Grünewald, 2011). The volume of people and efforts in and out of the country, on top of the devastation and losses, must have felt confusing and overwhelming. Who coordinates such an influx of help? Although governments should take on the responsibility of coordination, the reality is that not all can do so. In Haiti, many government and nonprofit personnel died in the event, creating a leadership vacuum.

International disaster humanitarian Fred Cuny's lifetime of experience led him to observe that how an organization makes decisions – with or without local input – will determine if the effort succeeds or fails. His experience suggests that "any effort without the full participation of the disaster victims misses the opportunity to increase the people's ability to make choices and to help them attain self-confidence in decision making" (Cuny, 1983, p. 92). Clearly, a paternalistic attitude, where outsiders make the choices, will not have much success. Rather, partnering from the grass roots should always be a first step.

In its humanitarian efforts, *Doctors Without Borders* follows ethical principles designed to embrace local ways. Based on their significant experience, they recommend that those new to a humanitarian relief setting first take time to observe, noting that seemingly unorthodox or ineffective practices may be in place for good reasons. Newly arrived volunteers are advised to take notes on things they might wonder about or think need to change, but wait to offer alternative procedures or processes until they understand the context for existing practices better. In short, patience and diligence through the process of accompaniment is recommended by those well-experienced in international humanitarian relief.

Waiting may seem to take forever, with volunteers and donors pressuring an agency to get in to an area and do something. Yet, unless pre-existing relationships are in place or teams with credentials deploy in an orchestrated effort, establishing partnerships is what should be done *before* arriving. Working with locals may require establishing partnerships with a range of actors – governmental and non-governmental organizations, international NGOs, faith-based organizations, the military, emergent groups, and survivors. Each actor will embrace a different way of doing things. Being able to understand their culture, and the context in which they follow their culture, will be key to success as it informs how organizations operate, interact, and collaborate. To illustrate, a medical team from Stanford University Hospital and Columbia University Medical Center traveled into Haiti after the earthquake along with literally hundreds of other health care agencies (Auerbach et al., 2010). After three days of struggling to support hundreds of patients, a U.S. infantry regiment arrived to provide security and perimeter access, and then supported efforts with medics and supplies. A naval hospital ship arrived within a week and the regiment transported patients offshore for surgeries and high-level care. Their biggest challenge? Communications and organizational structures, both of which are informed by culture. Military language, for example, may include many unfamiliar acronyms – just as the language of medical care could result in misunderstandings.

Beyond the essential role of culture, organizations need to form lasting relationships on the long journey to recovery. Short duration trips or relationships fail to help volunteers understand local context including the impact of their presence and own behaviors, like snapping selfies for social media (Van Hovoing, Wallis, Docrat, & De Vries, 2010; Welty & Bolton, 2017). At the core of the problem is accountability for one's actions, both during the volunteer trip and for what volunteer efforts have left behind including loss of jobs, inabilities to sustain what has been initiated, or feeling abandoned. Emphasizing learning as part of service can address such shortcomings, as can developing longer-term and thus meaningful and accountable efforts. Volunteers then return home as advocates for the people, culture, and contexts in which they have served and leave people feeling better about themselves and their situation.

However, a realistic question that merits attention is this: given that disasters undermine governmental and NGO capacities to respond, should outsiders truly rely on locals? Essentially, doing so means involving sometimes traumatized survivors in humanitarian and international disaster volunteer work. Research suggests that such involvement may actually provide therapeutic benefits (Karanci & Acaturk, 2005). In short, involve and rely on locals while being sensitive to culture, context, and impacts of volunteer efforts from response through recovery time periods. International disaster volunteering efforts are far more likely to succeed for those who do so.

"Invited to gumbo"

Survivors of disaster know the difference between authentic, genuine volunteers and those who arrive for a photo opportunity. They form meaningful relationships with people who come for the right reasons and remember

PHOTO 8.2
Residents of the Lower Ninth Ward protest hurricane Katrina disaster tourism.

them long after those volunteers have moved on to the next disaster project. In a study of volunteer efforts after hurricane Katrina, I found that people whose homes were repaired or rebuilt bonded even with those who came for only a week. They missed their volunteers after they left, having shared a time in their lives when significant help and healing took place after a formidable trauma (Phillips, 2014, 2015). The meaningful connections they forged with volunteers contrasted sharply with busloads of tourists who drove through the devastated areas taking pictures then leaving. Locals resented the intrusive ways in what such behaviors objectified their tragedy for social media postings (see Photo 8.2). People in Haiti also reported feeling objectified through photos taken by outsiders, feeling that turned them into anonymous others. Called "voluntourists," such a divisive us-versus-them approach hurts local survivors (Welty & Bolton, 2017).

I interviewed many Katrina beneficiaries, by sitting in their homes and listening as they offered stories about their experiences. They pulled out photo albums, pointed to pictures on the walls, and talked about phone calls, emails, and visits from their volunteers long after the home had been rebuilt. Although I asked about the quality of the rebuilt homes, it became quickly clear that the volunteer experience was not about the house. Survivors wanted to talk more about heartwarming volunteer encounters that had transformed their lives and the people who put their own lives and work aside to help. As one person said to me about Canadian volunteers: "I don't think there was anyone left in Saskatchewan." They told joyful stories of sharing their culture, particularly local cuisine that represented their cultural heart.

In Louisiana's bayou communities, many people shrimp or fish for a living. One particularly relished dish is gumbo, a stew made from a stock thickened by roux then enriched with the "holy trinity" of celery, onion, and green bell peppers. Local "Cajun" spices infuse and build familiar flavor profiles, as the gumbo simmers with okra, crabmeat, and shrimp which is then served over rice. Gumbo represents the heart of Louisiana Cajun cooking, a tangible representation of people's connection to places imbued with meaning for them (Chamlee-Wright & Storr, 2009). Gumbo also symbolizes the array of cultures that make up a complex mix of people along the coast from New Orleans neighborhoods into the isolated bayous. The music of New Orleans has often been referred to symbolically as gumbo, with a mix of jazz, blues, zydeco, Cajun, rhythm and blues, rock and roll, soul, and gospel (Lichtenstein & Dankner, 1993). When people share gumbo, they share themselves.

In a mobile home along the coast, I talked to a woman where volunteers were repairing her home after Katrina. Years had passed since the storm, with most volunteer groups having left the area a long time ago. Her family, shrimpers all, had faced a significant economic decline and the impact of a major offshore oil spill in the years following the storm. As we talked, she peeled tiny shrimp – made small because of the oil spill damage done to the ecosystem. She told me about the Katrina volunteers who kept coming, year after year, into the region where she lived. Her volunteers listened as she and her neighbors told stories about the storm, their way of life, and their losses. She talked about the volunteers who kept coming until they reached her house. They did not look for photo opportunities. They did not take money and then disappear as unscrupulous contractors had done. They returned phone calls and became good stewards of her limited funds. Her volunteers called back to check and see how things were going, even well after one team had been replaced by the next. She told me her door was always open to those volunteers, to anyone associated with their organization. She smiled at me as she passed a bowl and said "everyone is invited to help, but not everyone is invited to gumbo."

Conclusion

This chapter is about how to be invited to gumbo (see Box 8.1). Strategies to develop meaningful and effective relationships rely on understanding the culture and context in which a disaster has occurred. Culture includes understanding the norms, values, language, and symbols of a given location which can be quite diverse. Norms include folkways and mores, or behavioral guidelines ranging from informal interaction to prohibited behaviors. Following norms enables outsiders to fit in and to make traumatized survivors more comfortable with strangers. Values guide outsiders to know what

Box 8.1 Strategies for becoming culturally competent

Pre-departure

- Learn about the culture of the host community. Do so by:
  - Inviting people from that culture to speak to your team.
  - Asking a local professor to offer insights to your team.
  - Learning about the social, political, economic and environmental structures and issues of the host country.
  - Developing an appreciation of music, art, poetry, literature, and other intangible aspects of people's culture. Imagine, for example, the value of art or music therapy within a refugee camp when working with children – or the familiar comfort of a local hymn, folk song, or dance tune. While time might not permit such cultural enrichment during emergency response times, recoveries can linger for years or even a decade, providing ample opportunity to fit in to the local cultures more effectively.
- Learn verbal forms of communication.
  - Acquire the basics: hello, goodbye, please, thank you, yes, no, and the ways in which people use language to offer respect and enhance dignity.
  - Practice greeting people appropriately.
- Learn non-verbal forms of communication.
  - Spend time understand the ways in which people dress and the relationship of clothing to faith, environment, gender, social class and other influences on personal appearance.
  - Learn how people use personal space.
  - Practice culturally appropriate handshakes and eye contact.
- How do people organize meetings? Where are meetings held? Who is in charge? How are decisions made?

In-Country

- Ask hosts about their culture. Tell them it is ok to offer insights, advice, and even corrections for inappropriate behaviors
- Hold nightly debriefing and discussion sessions with your team. Be sure to include insights on cultural interactions as a means to learn from the team.
- Encourage people to journal or engage in creative ways to explore the culture such as photographs (with permission), drawings, poetry, or other ways to express themselves.
- Participate in local cultural traditions when invited and share yours when appropriate.
  - Try local foods and express interest in how foods are made and where they come from.
  - Listen to stories from the local culture including ones about the disaster.
  - Visit locally meaningful places like locations for worship, statues, parks, and public spaces that have survived.
  - Go to someone's home if invited and take something appropriate as a thank you such as flowers or a symbolic gift from your own culture.
  - Be a diplomat.
- Reflect on reactions to new things and think through the potential basis for any negative reactions. Work with the team to try and understand from the local point of view.

is considered desirable and undesirable from clothing to food to physical contact. Language helps people fit in and demonstrates respect for locals who may be struggling to understand many newly arriving language groups. By understanding the meaning of symbols, we can interact with people more effectively and respect their way of life more fully. Context includes being sensitive to time, place, and circumstances that influence how an international disaster team may be able to arrive and help or not. Overall, understanding and valuing local culture and context enables an outside organization or team to be more successful and to reduce any unintentional harm that could ensue. Be invited to gumbo.

References

Auerbach, P., Norris, R., Menon, A., Brown, I., Kuah, S., Schwieger, J., . . . Lawry, L. (2010). Civil-military collaboration in the initial medical response to the earthquake in Haiti. *The New England Journal of Medicine, e32*(2), 1–4.

Barber, R. (2009). The responsibility to protect the survivors of natural disaster: Cyclone Nargis, a case study. *Journal of Conflict & Security Law, 14*(1), 3–34.

Benjamin, E., Bassily-Marcus, A. M., Babu, E., Silver, L., & Marin, M. (2011). Principles and practice of disaster relief: Lessons from Haiti. *Mount Sinai Journal of Medicine, 78*, 306–318.

Center for Disease Control & Prevention. (2005). Fatal injuries among volunteer workers – Untied States, 1993-2002. Retrieved from https://www.cdc.gov/mmwr/preview/mmwrhtml/mm5430a2.htm, Accessed 21.12.17.

Chamlee-Wright, E., & Storr, V. (2009). "There's no place like New Orleans": Sense of place and community recovery in the Ninth Ward after hurricane Katrina. *Journal of Urban Affairs, 31* (5), 615–634.

Cunningham, T., & Sesay, A. (2017). The triple menace in volunteer international aid work: Three harmful pitfalls. *Journal of Emergency Nursing, 43*(5), 478–481.

Cuny, F. (1983). *Disasters and development.* Dallas, TX: Oxford University Press.

Davidson, C., Johnson, C., Lizarralde, G., Dikmen, N., & Sliwinski, A. (2006). Truth and myths about community participation in post-disaster housing projects. *Habitat International, 31*(1), 100–115.

Farmer, P. (2011). *Haiti: after the earthquake.* NY: Perseus Books.

Fisher, S. (2010). Violence against women and natural disasters: Findings from post-tsunami Sri Lanka. *Violence Against Women, 16*(8), 902–918.

Frerichs, R., Keim, P., Barrais, R., & Piarroux, R. (2012). Nepalese origin of cholera epidemic in Haiti. *Clinical Microbiology and Infection, 18*(6), E158–E163.

Hampson, J., Crea, T., Calvo, R., & Álvarez, F. (2014). The value of accompaniment. *Forced immigration review.* Available from http://www.fmreview.org/faith/hampson-crea-calvo-alvarez.html, Accessed 20.01.18.

Handmer, J., & Nalau. (2013). Is relocation transformation? http://research-hub.griffith.edu.au/display/nb44819abc2abb41486d9f5b324cc5921, Accessed 05.01.16.

Hull, C. (2009). Tale of war and peace in the 2004 tsunami. Available from https://www.reuters.com/article/us-tsunami-anniversary-conflict/tale-of-war-and-peace-in-the-2004-tsunami-idUSTRE5BH01O20091218, Accessed 21.11.17.

Jasparro, C., & Taylor, J. (2011). Transnational geopolitical competition and natural disasters: Lessons from the Indian Ocean Tsunami. In P. Karan, & S. Subbiah (Eds.), *The Indian Ocean Tsunami* (pp. 283–299). Lexington KY: University of Kentucky Press.

Jobe, K. (2011). Disaster relief in post-earthquake Haiti: Unintended consequences of humanitarian volunteerism. *Travel Medicine and Infectious Disease, 9*, 1–5.

Kanamori, J., Uchiyama, B., Hirakata, Y., Chiba, T., Okuda, M., & Kaku, M. (2013). Lessons learned from a tuberculosis contact investigation associated with a disaster volunteer after the 2011 Great East Japan earthquake. *The American Thoracic Society, 187*, 1278–1279.

Karanci, N., & Acaturk, C. (2005). Post-traumatic growth among Marmara earthquake survivors involved in disaster preparedness as volunteers. *Traumatology, 11*(4), 307–323.

Lichtenstein, G., & Dankner, L. (1993). *Musical gumbo: The music of New Orleans.* NY: W.W. Norton & Company.

Merchant, R., Leigh, J., & Lurie, N. (2010). Health care volunteers and disaster response – first, be prepared. *The New England Journal of Medicine, 362*(10), 872–873.

Miller, M., & S. Garrett. (2009). Improving disaster volunteer safety through data collection and skills matching. In *Proceedings of the 2009 industrial engineering research conference.*

Pan American Health Organization/World Health Organization. (2017). Changing health risks and interventions due to earthquakes and hurricanes. Retrieved from http://www.paho.org/hq/index.php?option = com_content&view = article&id = 13702&Itemid = 135&lang = en, Accessed 20.12.17.

Goyet, C., Sarmiento, J., & Grünewald, F. (2011). *Health response to the earthquake in Haiti January 2010: Lessons to be learned for the next massive sudden-onset disaster.* Pan American Health Organization/World Health Organization.

Picou, J. S. (2000). The "talking circle" as sociological practice: Cultural transformations of chronic disaster impacts. *Sociological Practice: A Journal of Clinical and Applied Sociology, 2*(2), 77–97.

Phillips, Brenda. (2014). *Mennonite Disaster Service: Building a therapeutic community after the Gulf Coast storms.* Lanham, MD: Lexington Books.

Phillips, B., & Thompson, M. (2013). Religion, faith and faith-based Organizations. In D. Thomas, B. Phillips, W. Lovekamp, & A. Fothergill (Eds.), *Social vulnerability to disasters* (pp. 341–366). Boca Raton, FL: CRC Press.

Phillips, B., Neal, D., Wikle, T., Subanthore, A., & Hyrapiet, S. (2008). Mass fatality management after the Indian Ocean tsunami. *Disaster Prevention and Management, 17*(5), 681–697.

Phillips, B., Stukes, P., & Jenkins, P. (2012). Freedom Hill is not for sale and neither is the Lower Ninth Ward. *Journal of Black Studies, 43*(4), 405–426.

Santos-Hernández, J., & Morrow, B. (2013). Language and literacy. In D. Thomas, et al. (Eds.), *Social Vulnerability to Disaster* (pp. 265–280). Boca Raton, FL: CRC Press.

Sayeed, A. (2009). Victims of earthquake and patriarchy: The 2005 Pakistan earthquake. In E. Enarson, & P. Chakrabarti (Eds.), *Women, Gender And Disaster: Global Issues And Initiatives* (pp. 142–151). Los Angeles: Sage.

Silove, D., & Zwi, A. (2005). Translating compassion into psychosocial aid after the tsunami. *The Lancet, 365*, 269–271.

Stover, E., & Vinck, P. (2008). Cyclone Nargis and the politics or relief and reconstruction aid in Burma (Myanmar). *Journal of the American Medical Association, 300*(6), 729–731.

Sumner, W. G. (1902). *Folkways: a study of the sociological importance of usages, manners, customs, mores, and morals.* NY: The Anhenaeum Press.

Takahashi, K., Kodama, M., & Kanda, H. (2013). Call for action for setting up an infectious disease control action plan for disaster area activities: Learning from the experience of checking suffering volunteers in the field after the Great East Japan earthquake. *Bioscience Trends, 7*(6), 294–295.

Van Hovoing, D., Wallis, L., Docrat, F., & De Vries, S. (2010). Haiti disaster tourism – a medical shame. *Prehospital and Disaster Medicine, 25*(3), 201–202.

Welty, E., & Bolton, M. (2017). The role of short term volunteers in responding to humanitarian crises: Lessons from the 2010 Haiti earthquake. In M. Holenweger, et al. (Eds.), *Leadership in extreme situations* (pp. 115–130). Advanced Sciences and Technologies for Security Applications. Available from https://doi.org/10.1007/978-3-319-55059-6_7.

Wickramage, K. (2006). Sri Lanka's post-tsunami psychosocial playground: Lessons for future psychosocial programming and interventions following disasters. *Intervention, 4*(2), 167–172.

CHAPTER 9

The benefits of volunteering

Introduction

Volunteers make a difference and disaster volunteers turn matters around after the worst day of people's lives. As introduced in Chapter 1, volunteer labor produces numerous benefits, and not only for the people and communities they serve. In this chapter, we look at the ways in which volunteering produces useful benefits for volunteers, survivors, emergency managers and volunteer coordinators, and affected communities. This chapter relies on evidence-based best practices supplemented by my own research-generated observations.

The benefits of disaster volunteering

Research documents some of the benefits of volunteering and, while not necessarily specific to disaster contexts, should be transferable to those who go into communities harmed by natural disasters, technological accidents, or terrorism. In general, volunteering produces a number of benefits for those who serve, ranging from physical and psychological health effects to gains in work experience, personal and professional development. Life satisfaction overall also appears to increase (Thoits & Hewitt, 2001; van Willigen, Bob Edwards, Terri Edwards, & Hessee, 2002). In this section, readers will learn more about benefits for all concerned: those who serve, survivors, emergency and volunteer managers, and communities.

For those who serve

Studies suggest that a possibly wide array of benefits may accrue to volunteers in general, and to those who serve in disasters as well. The benefits of general volunteering appear to include feeling better physically and psychologically (Stukas, Hoye, Nicholson, Brown, & Aisbett, 2016) quite possibly arising from a heightened sense of social connectedness and interactions.

Disaster Volunteers. DOI: https://doi.org/10.1016/B978-0-12-813846-5.00009-5

Opportunities to serve also generate personal and professional side benefits. Personal benefits may include new knowledge about how to care for others, as well as an enhanced ability to work on one's home after doing disaster repairs. Disaster volunteers also report a more informed understanding of disasters, which can produce insights that produce heightened safety concerns at home. Professionally, people may acquire new knowledge and skills suitable for the workplace. Benefits may also accumulate for people from various demographics, if people are not overlooked. This section reviews those benefits and some related concerns.

Some studies report that physical benefits amass for volunteers. Older volunteers in particular have indicated a health increase and an overall sense of wellness (Piliavin & Siegl, 2007; van Willigen et al., 2002; Wilson & Musick, 1997). Care should be taken in assuming such increases though, because it may be that healthier volunteers turn out which could influence those findings, including for older volunteers. It may also be that the physical movement involved in volunteering will bump up one's health and fitness levels or just make people feel better. Volunteering also seems to protect against cognitive aging by increasing physical, social, and cognitive activity thus improving older volunteers' overall functioning and sense of feeling better (Guiney & Machado, 2018). Volunteering, and the effort involved, is certainly consistent with medical advice to keep moving as we age.

Even more studies indicate that volunteers feel a psychological benefit or mental health boost from their efforts (e.g., see Piliavin & Siegl, 2007; Thoits & Hewitt, 2001). The reason for that boost may come from a variety of sources. Volunteers who are already mentally healthy may be more ready to volunteer and thus report higher mental health (Santini et al., 2019). It could also be that by volunteering, people move away from familiar stress and upset and get their minds off of their problems. The social nature of volunteering may also be part of the mental health boost, because interacting with other people may reduce isolation, affirm one's sense of self-esteem, and build new friendships (e.g., see Thoits & Hewitt, 2001). Volunteer sites seem to provide a place where people feel they belong and a space in which to regenerate, not unlike taking time off from work or getting away from one's worries (Berg & Johansen, 2017). Studies of psychological benefits specific to disaster settings remain meager, but one found evidence that people dealing with difficult situations like divorce or depression felt better, perhaps because disaster sites generate new perspectives on one's own concerns (Phillips, 2014).

People may feel better because they acquire new skill sets that make them feel more confident or capable. Female volunteer firefighters in Australia reported feeling more confident, coupled with a strong sense of achievement after they turned out to fight bushfires (Branch-Smith & Pooley, 2010). Other skills may be acquired as well. For example, volunteering has the potential to benefit careers by affording opportunities to engage in and enhance teamwork capabilities. Dealing with a difficult rebuilding site will also require creative problem-solving. Volunteers who lean into leadership positions may also find themselves interacting with people outside the disaster organization, like building code officials or long term recovery groups. They will need to learn how to discern and navigate other's perspectives and requirements to succeed in their volunteer effort. Doing so will require communication and collaboration to produce results. All of these volunteer activities will build someone's repertoire of work and people skills suitable to building a resume and advancing one's careers.

Feeling more confident may very well extend to one's personal life as well. Working on a disaster reconstruction site may offer chances to learn home management skills like painting, sawing, hammering and abilities to install, mud, and tape drywall or shingle roofs. Disaster work areas also require people who can cook or assist the feeding effort including how to address a range of nutritional needs, allergies, and dietary requirements sometimes for dozens of fellow volunteers. Such personal growth is a clear benefit of volunteering, as one study of volunteers over 65 in Norway confirmed (Berg & Johansen, 2017). Personal growth may also result in spillover when volunteers return home. For example, learning to cook for multiple dietary needs on a large volunteer site can transfer into feeding the hungry at local soup kitchens once returning home. Such "spillover," from one kind of volunteering into another, then benefits one's home community.

A disaster site also allows volunteers to learn about the powerful impacts of nature or human intervention. Volunteers report being stunned at what a tornado, hurricane, or earthquake can do and marvel at the damage extending for miles. Disaster site experience thus appears to sensitize people to the power of nature, the trauma that people experience, and the magnitude of loss (Phillips, 2014). The visual impact from serving on a disaster site may spur volunteers to think through their own safety and that of their community. Returning with new insights, they may look at area hazards and prepare their own households to be more resilient. Certainly, returning volunteers would then be suitable for recruitment into local emergency response teams and volunteer groups. Emergency and volunteer managers may want to follow up when local teams return from a disaster to recruit them into local efforts, thus triggering the spillover effect more directly.

Disaster sites also give people the chance to build insights into why people want to live in harm's way. Volunteers going into areas that historically flood may initially wonder why people would stay there and face repetitive risks.

After meeting the people, learning about their way of life, and spending time interacting, a greater understanding and level of support may develop. That understanding may include deeper appreciation of other cultures and places. After hurricane Katrina, Canadian volunteers came into Louisiana and helped across a considerably diverse set of communities that characterized the U.S. Gulf Coast. They learned new foods like gumbo and po' boy sandwiches while sharing their own food preferences. Volunteer sites brought in local residents to their temporary housing for meals, music, and recreation. Their interaction produced even deeper sets of ties among former strangers who became friends across borders (Phillips, 2014), something that survivors report as well (see next section).

Those social ties matter to volunteers and likely helps to generate other benefits. As learned in previous chapters, volunteers can often be recruited from within social networks of people who share similar values. Working with like-minded believers strengthens one's own values and can deepen a sense of connection to faith or organization (Becker & Dhingra, 2001; Phillips, 2014). Volunteering facilitates that sense of fellowship, by providing a shared site in which social networks of people feel validated, build connections, and leave reaffirmed (Phillips, 2014; Santini et al., 2019).

Older volunteers seem especially likely to benefit. One study of Taiwanese volunteers over age 65 found that volunteering allowed them to rediscover a purpose in life after retiring from work (Ho, 2017). Given that older volunteers had more free time, volunteering provided a goal they could focus on to feel useful and to insure they continued to live a meaningful life. At younger ages, teenagers who volunteered in New Zealand said that serving gave them a "sense of purpose or control" as well as perspectives on the disaster and their own situations (Pine, Tarrant, Lyons, & Leathem, 2018, p. 376). A Polish researcher found similar results among adolescents after a flood, particularly more proactive coping (Bokszczanin, 2012). She recommended involving teens after an event coupled with appropriate monitoring and supervision to not overburden younger volunteers new to disaster scenes.

However, people with disabilities remain overlooked as volunteers and for disaster response (Bruce, 2006; National Council on Disability, 2009; Shandra, 2017). Such potential helpers want to participate but often face barriers, sometimes arising out unreliable assumptions, about what they can do. People with disabilities thus represent an "untapped supply of volunteer labor" as a result of "social exclusion" (Shandra, 2017, p. 207). Their contributions can be brought into play by having managers and organizations build partnerships with disability organizations and advocates, offering cross-training with those organizations, and viewing them as active partners in disaster response. Organizations can also level the playing field by first viewing people with disabilities as enabled volunteers who can self-determine their preferred ways to help and by creating a space where all volunteers can communicate despite hearing differences, mobility and sight challenges, or cognitive functioning. Given their higher levels of volunteering than people without disabilities, offering volunteer opportunities broadens the benefits to all and enriches the pool of volunteer labor. Imagine, for example, rebuilding the home of a senior citizen who uses a wheelchair. A volunteer with a disability can offer insights into ramps, counter heights, door levers instead of knobs, floor plans, toilet height, accessible showers, and specialized smoke detectors. They will be far more knowledgeable about ideas from universal design principles that enable access not only for survivors with disabilities but for everyone.

Similarly, managers and organizations often overlook low-income volunteers who are often seen as aid recipients rather than helpers (Benenson & Stagg, 2016). But low-income volunteers also bring considerable, often under-used assets to disaster settings. Such social capital can enable new ways of thinking, new connections, and new approaches to volunteer organizations (Isham, Kolodinsky, & Kimberly, 2006). People who are low-income bring savvy ways to address economic challenges, creative strategies for survival, and ties to other people and networks. By reversing one's perspective on who can and cannot help, the same benefits that higher income volunteers experience can also become available to those at lower income levels – and bridge greater understanding between sometimes distant groups of people. This idea, called the contact hypothesis (usually used to address and reduce racial and ethnic stereotyping), means that when people come into interaction with each other, prior assumptions and misperceptions become transformed as authentic and meaningful relationships develop (Hewstone & Swart, 2011). Emergency managers and volunteer coordinators can use the contact hypothesis to create more bridging social capital, and truly involve a fuller range of volunteers to benefit their communities and the volunteers as well.

For those who are served and those who survived

When considering the beneficiaries of disaster volunteerism, most reports focus on what they receive: meals served, donations given, clean up kits offered, and homes that are transformed, repaired, or rebuilt. Annual reports document the numbers of each to tally the tangible results of helping after a disaster. Intangible results also result from disaster volunteering, but this kind of qualitative data remains largely absent from reports and research. Nonetheless, people do experience meaningful, human connections that restore those affected and enable survivors to move on. Barton (1970) observed this therapeutic effect when volunteers arrive (as described in previous chapters). Such an uplifting of the human spirit often stands in contrast to the more bureaucratic response of government. Government and similar bureaucratic organizations, despite often trying their best to serve survivors, often become the target of people's pain and disappointment. People rarely experience government or bureaucracy in a "warm and fuzzy" way, probably because bureaucracies often use a one-size-fits-all approach to serve many people quickly and equitably. Survivors report frustrations with trying to secure the aid they need in a timely way because of confusing paperwork and aid denials, troubles reaching the right people in the government office, waiting for phone calls to be returned, standing in line or competing for resources, and the timeliness of receiving needed aid. They compete for resources, often giving up in sheer frustration.

Then, volunteers appear. Beneficiaries report that it feels like the cavalry has arrived and that they can begin to hope again (Phillips, 2014). Volunteers and their experienced organizations can often do what government and bureaucracies cannot, including addressing unmet, unique, or newly emerging needs. They can bend and flex to listen to what survivors need, adapting their own ways of doing things and even transforming what they had planned to do. In one instance, a survivor repeatedly declined having a disaster organization build a house for him after viewing several plans (Phillips, 2014). The volunteers decided not to give up, and asked him what he really wanted in a home. When he explained his simple and straightforward needs, they redesigned and downsized their building plans including a special area for his beloved pets. To everyone's surprise, he agreed to the plans. Perhaps the most poignant memory came from a woman whose spouse had died shortly after their home had been repaired. She said that while he was in the hospital moving into and out of consciousness, she heard him whispering people's names. As she leaned in close, she realized he was calling out the names of the volunteers who had worked on their homes. Another woman I met said that she greeted every volunteer team with the same words every morning: "my angels have arrived!" It appears that peace of mind is what volunteers bring to survivors (Unruh, 2017).

Disaster beneficiaries also report making new friends, with whom they try to stay in touch. After Katrina, I visited with dozens of survivors in homes rebuilt by volunteers. They frequently pointed out photos of volunteers on the walls of their homes or proudly produced entire photo albums. They had lists of their volunteers' emails and addresses and exchanged cards at holiday times. Some volunteers came back to visit their homeowners, staying for meals of oysters and boiled shrimp bringing their own confections to share. In a few cases, survivors visited their volunteers including a group of Louisiana Cajun French homeowners who toured Amish homes in

Pennsylvania. For several, lifelong friendships resulted with people going on joint vacations or spending time in each other's homes (Phillips, 2014).

Survivors, from homeowners to community leaders, also come to trust outsiders whom they distinguish from those who show up for a photo opportunity. People I spoke with at the community level said that they knew who they could trust to get the job done correctly, which organization would follow their local building codes, and treat their people with respect. Learning to trust outsiders also included survivors overcoming fear of fraudulent contractors in order to entrust their savings or government grants to people in organizations they had never heard of. They told me that even when they could not get through to the government 1–800 numbers, they could always reach the faithful volunteers associated with disaster organizations. Multiple community leaders told me, with surprise, that volunteers and their organizations always answered the phone or called back soon. One community leader told me that "God's area code" was the same as the one for the disaster organization rebuilding homes in their community. This reliable and very human connection, during a deeply personal and community-wide ordeal, made a difference and helped to produce that desirable therapeutic effect (Barton, 1970).

Survivors also told me about deepening their own cross-cultural connections and understanding. They learned about new food, asked questions about unfamiliar faiths, and acquired new vocabulary. They shared from their own culture as well by making or buying food for volunteers, praying for those who were helping them, teaching outsiders about their environment or language, and visiting volunteers in their camps and temporary service sites. Survivors and volunteers went on an unexpected journey together, from despair to hope and from homeless to home again. Volunteers enabled families to reunite, often from significant displacement, and made their families whole again (Phillips, 2014). Volunteers restored their faith in humankind.

In numerous instances, beneficiaries went on to volunteer in their own communities or on other disaster sites, suggesting that a spillover or "pay-it-forward" spirit may result for them as well (Phillips, 2014, 2015). In some cases, disaster survivors showed up at a disaster site in another location, sometimes traveling hundreds of miles to clean up and repair the homes of others. Those who could donate financially often did so or found another means to repay like praying for the volunteers or sweeping the floor every night after volunteers left. Survivors also found ways to serve closer to home (Phillips, 2014). One woman I met, whose home was destroyed by hurricane Ike in Texas, then went on to volunteer at a local food bank. Another man, who had been self-isolated and dealing with significant health concerns, cleared land behind his new home built by disaster volunteers and created a multi-acre community garden. In doing so, he lost weight and reduced his medications – and began delivering fresh vegetables around his community. Volunteers transformed survivors' lives, perspectives, and futures.

For emergency managers and volunteer coordinators

Facing a massive pile of destruction that lingers for miles can certainly be daunting to even the most experienced emergency managers, volunteer coordinators, elected officials, and administrators. The expertise of emergency managers will be needed for many tasks after a disaster occurs from running the Emergency Operations Center to providing guidance to local politicians or working with the media. They will need to address a full range of challenges from getting emergency vehicles through damaged areas, supporting search and rescue efforts, assessing damage, setting up contracts for debris removal, and much more. Volunteer coordinators will have their hands full too, as they lean into the tasks identified by preliminary damage and needs assessments. Elected officials and community leaders will run or participate in countless, exhausting meetings. With an effective volunteer management plan, such professionals and officials can direct help appropriately and attend to other pressing matters. Those volunteers may also save the broader community and individual residents a considerable amount of money.

Monetary benefits specifically for emergency managers can be challenging to measure because the benefits can be rather indirect (Bachner, Seebauer, Pfurtscheller, & Brucker, 2016). One study in Austria found financial benefits for both emergency managers and the broader community. For the emergency manager, the costs of providing emergency services decreased because unpaid labor augmented budgets. More broadly, economic benefits resulted when volunteer efforts resulted in households being protected from floods. More indirect macro-economic gains also accumulated due to the perceived mental health and social capital effects of reduced disaster impacts (Bachner et al., 2016).

Another benefit for both emergency managers and volunteer coordinators is the opportunity to test volunteer management plans in a real disaster situation. Although tabletop and full field exercises can test the plan to a degree (Kim, 2014; Sinclair, Doyle, Johnston, & Paton, 2012), an actual event will reveal the strengths and weaknesses of any plan as well as places for improvement. Thus, a real disaster generates a latent benefit of being able to see if the plan works and to consider what needs to be revised. No plan is perfect, and the reality of a lived experience sheds light into what needs improvement, typically communication, collaboration, information exchange, management procedures, documentation, adaptability and capacity.

Testing those plans and anticipating future uses thus represents a perhaps unanticipated benefit for managers. As Chapter 10 will reveal, newer and even more challenging disasters lie ahead. As disasters continue to evolve and change, and to produce new and larger impacts, volunteer efforts will need to adapt to both traditional and emerging events like pandemics, geomagnetic storms, climate change, and terror attacks. Emergency managers will benefit from harnessing such a fuller range of skills and abilities and volunteer coordinators will benefit from being able to match individual turnout with disaster-driven needs. For example, The Standby Task Force that has involved digital crisis mapping volunteers in 80 countries represents one of the more recent adaptations that has brought in technology skills to volunteer efforts (Whittaker, McLennan, & Handmer, 2015). The good news is that people's skills and volunteer interests have been gravitating toward such needs, which produces benefits for managers willing to use them.

Ideally, the result of pre-disaster planning coupled with effective implementation will result in reduced stress for those tasked with managing the disaster and those who come to help. If things go well, officials will worry less about their community, neighbors, friends and even their own family. In the exhausting aftermath of a disaster, stress reduction will be invaluable for the professionals helping to put things back together. To feel confident these benefits will result, pre-planning is key to know what will happen and who will step up.

For communities that are served

When emergency managers and volunteer coordinators handle volunteer labor well – and when volunteers and disaster organizations accept their own responsibilities – the impacts to the broader community can be considerable. The speed of recovery can increase if a carefully coordinated effort falls into place. The number of displaced residents can be reduced and people can get back to work from individual jobs to small businesses to large corporations. Earning money means that mortgages and rent can be paid, banks can survive, and secondary businesses like gas stations and dry cleaners will thrive again. Residents returning home means that participation in civic, social, and political activities can be restored, which benefits many.

The example from Joplin, Missouri, mentioned previously shows that communities receive significant economic benefits in addressing the costs of disasters to their people (Abramson & Culp, 2013; Seeley, 2014). Enumerating those numbers can produce significant benefits for managers. The number of volunteers who went to Katrina exceeded half a million people in the first year after the hurricane, a phenomenal amount of free labor (Corporation for National & Community Service, 2006). Salvation Army volunteers served over five million hot meals, saving money and time for local residents who could focus on picking up after the storm. Episcopal Relief and Development gave free medical services to 22,000 people, supporting hundreds of medical professionals whose offices and homes had flooded. Travelers Aid helped 18,725 people get to safe shelter or reunite with loved ones. Angel Flights provided free air travel to disaster organizations assessing where to set up relief operations. Noah's Wish rescued 1900 dogs, cats, and other pets and arranged for medical and foster care. Volunteers for KaBOOM! agreed to rebuild 100 playgrounds in damaged areas. Clearly, communities derive considerable and wide-ranging savings from volunteer help. In addition to disaster-specific work, volunteers and their organizations spend money on resources, tools, construction supplies, housing, food, gas, and entertainment while in the affected area.

Often, the most visible benefit that can be seen comes from efforts that remove muddy debris, earthquake rubble, or charred buildings and restore familiar, clean environments to communities. The influx of people, though, brings other less obvious benefits. Disasters have the potential to undermine the tax base for a community, which can threaten programs and needs for years to come. Volunteers also pay taxes on many things they buy. They help re-open coffee shops and restaurants, fuel recreational, sports, and tourist venues, stay in hotels, beds and breakfasts, and shop in local stores. Communities also recognize this, and often provide places to stay, food and beverages, free utilities, transportation, medical care, and discounts to their volunteers. In a way, the influx of volunteers allows communities to demonstrate their appreciation by caring for those who come to help address wide-reaching, post-disaster needs.

Volunteers and disaster organizations also bring a wide range of expertise and interests that can holistically address disaster impacts beyond physical damage. Floods tumble trees into rivers, undermining riparian systems essential to ecosystem integrity. Volunteers can help with cleanup and restoration. Locally important heritage sites can be cleaned and repaired with the right expertise and supervision, so that tourism can rebound or local culture can be celebrated. After Katrina, some friends and I went to the Café du Monde in New Orleans the day it re-opened in the city's historic French Quarter frequented by tourists and vital to the area's economy. We all savored the beignet donuts and chicory flavored coffee distinctive of this unique and diverse city. The wait staff were relieved to have their jobs back and fall into a familiar routine.

But volunteer labor addresses more than the physical damage or economic recovery. Volunteer efforts often embrace those facing the hardest time recovering from a disaster. From a community-level perspective, that means that the area will remain enriched and diversified by income, age, race and ethnicity, dis/ability, culture, and family types. Low-income families and single mothers will be able to remain in the community near kin and friends, at their same jobs, and near affordable child care. People with disabilities will be able to return to their social and health care networks, from veterans facing mobility challenges to seniors who prefer trusted and familiar medical providers. The area will remain diverse also by culture, by helping urban areas like Haiti's capitol where people speak multiple languages and celebrate multiple nation-starting origins. Volunteers help restore the core and soul of a community, like the musician's village built in New Orleans with funding from famous musicians and volunteers through Habitat for Humanity (see Photo 9.1). A few years after hurricane Katrina, a local friend took me to visit the musicians' village under construction. As we sat there in her car, someone played a guitar. She turned to me, with tears in her eyes and said "the music is back." Volunteering brings the heart of a community back home and provides tremendously valuable healing to people who have endured damage to more than their homes.

After the Kobe, Japan earthquake, nursing teachers and students went into the homes of survivors, many of whom were living in temporary housing called "kasetu" on or near their campus (Kako & Ikeda, 2009). The earthquake seemed to increase health problems due to stress. Sadly, in the first

PHOTO 9.1

Volunteers built a "Musician's Village" elevated to resist future flooding and reflective of new Orleans' culture. *Courtesy: Brenda Phillips.*

year after the earthquake, 157 "kodokushi" or deaths alone were noted, with the majority (115) among men. In addition, many elderly residents faced displacement from their homes into new locations away from valued social networks. A primary consequence of the volunteer effort, just visiting with kasetu reidents, reduced people feeling isolated and alone. Simply spending time and listening to people reduces such isolation, which is a leading cause of suicide (Merchant, Leigh, & Lurie, 2010; Nakhaei et al., 2015; Quevillon, Gray, Erickson, Gonzalez, & Jacobs, 2016; Wang, Ip, & Chan, 2016). The broader community thus benefited when these Japanese volunteers, who served the displaced for 4.5 years, addressed unanticipated consequences of a natural disaster. Some volunteers continued their help in the residents' final, newly rebuilt location called "Happy Active Town Kobe" (Kako & Ikeda, 2009).

Clearly, benefits of all kinds arise as a consequence of people's prosocial behavior. The disaster context provides opportunities to acquire such benefits. Communities and organizations will want to keep those volunteers coming back during recovery and going on to a future disaster. As a means to keep the benefits discussed in this chapter continuing, the next section will address how to keep volunteers.

Volunteer retention

Volunteers come and go, but what makes them stay? Some show up for a short-lived time of help while others stay in for long term service. They may stay with a single organization for a lifetime or switch from one to another. Life circumstances, often linked to demographics, may mean they step out of volunteering and then return. They may prefer to do their work locally or at a distance or even through digital means. Retention and attrition represent the two concepts that address the dilemmas that disaster organizations, emergency managers, and volunteer managers face in preserving the valuable benefits that volunteers provide. Retention, which refers to keeping volunteers with you or your organization, can vary considerably – and can be influenced in a positive way. Attrition means people leave or drift way from a volunteer effort. Ideally, organizations, managers, and coordinators will want to do all they can to retain their volunteers. But, what works best to make that happen? Organizations should have a strategy to recruit their volunteers (see Chapter 4) and to retain them (Devaney, Kearns, Fives, & Canavan, 2015) (Fig. 9.1).

Just as the NVOAD Points of Consensus affirmed in Chapter 7, an emerging body of research suggests that relationships matter to volunteers. Thus, a first principle of retention for any organization or manager is to focus on

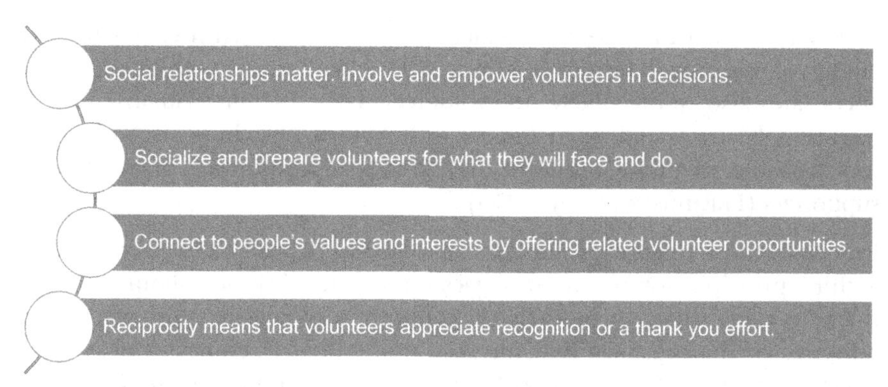

Social relationships matter. Involve and empower volunteers in decisions.
Socialize and prepare volunteers for what they will face and do.
Connect to people's values and interests by offering related volunteer opportunities.
Reciprocity means that volunteers appreciate recognition or a thank you effort.

FIGURE 9.1
Principles of volunteer retention.

relationships – yes, soft skills or people skills do matter. For example, one study in Australia surveyed 721 volunteers helping mostly health and community organizations (Walker, Accadia, & Costa, 2016). Two key findings point in a promising direction. The first is that organizations need to provide clear and visible support to volunteers in order to decrease attrition. Such an effort might include involving volunteers in organizational decisions and providing training. Organizations that empower volunteers as decision-makers seem to retain volunteers better (Farmer & Fedor, 1999). Surely, disaster context volunteers would appreciate the same degree of involvement.

The demographics of volunteers may matter as well (Lum & Lightfoot, 2005). Younger volunteers may value different kinds of experiences, interactions, and relationships with the organization. Today, over five very different generations are in the workplace – and their myriad interests and approaches are also in the volunteer sector. A second principle of retention then, is to address volunteers' interests and values. For example, the possibility of enhancing one's career speaks to younger volunteers (Erasmus & Morey, 2016). Organizations that want to retain younger volunteers may want to feature benefits associated with career pathways and work opportunities. Middle-aged volunteers may need alternatives to leaving family and home, such as using technology, providing administrative support from one's home or workplace, training other volunteers before they leave, arranging for recognition ceremonies when they return, and helping with fundraising. Older volunteers seem to value social functions, in that volunteering gives people – especially those who are retired – an opportunity to interact (Erasmus & Morey, 2016). Conversely, after retirement, people find that their interactions drop off and volunteering can fill that void. Organizations seeking to retain older volunteers may want to design and then feature social opportunities that enable them to continue interacting with like-minded friends who also enjoy volunteering (Erasmus & Morey, 2016). Volunteers with good experiences are also more likely to volunteer later in life (Adams, 1990). Interestingly, by attending to the social functions of helping, organizations may be able to turn volunteers into long-term committed organizational supporters (Erasmus & Morey, 2016).

Even with those interests, volunteers still need to be readied for service. Thus, a third principle for retention suggests that organizations should prepare their "workers" through providing a continuum of socialization experiences reflective of the organization's mission and culture (Devaney et al., 2015). Socialization is best described as the process of becoming a member of society or, in this case, of an organization. What are the organization's values and norms? What does the organization embrace as a design for interacting with both volunteers and beneficiaries of their efforts? Providing training offers one route to prepare volunteers to serve, which may be most important in the early stages of a volunteer's experience with the organization. Short courses, workshops, conferences, and events where volunteers can practice skills, share experiences, and build relationships offer some routes to socialize volunteers. Failure to properly prepare volunteers for what they will face may result in significant attrition. A small study of volunteers raising guide dogs in Australia found this, as volunteers realized the work was not what they expected and opted out before completing their service (Chur-Hansen, Werner, McGuiness, & Hazel, 2015). People may not fulfill their obligations and may not return if the organization does not ready them for their service.

A related concern is that volunteers tend to be less prepared for disaster contexts than professionals (Taylor & Frazer, 1981; Thompson, 1991; Thormar et al., 2010). Disaster-time mental health impacts can be negative, particularly when volunteers have not been properly trained or prepared before exposure to difficult scenes or settings (Thormar et al., 2010). Organizations can help retain their volunteers by training to be ready for hard situations involving acute exposure. Organizations should also arrange for debriefing, respite, and recreation, and by having volunteers stop out to avoid negative impacts (Thormar et al., 2010).

Volunteers also look for something reflective of a professional environment not unlike paid labor. Part of this comes from a fourth key retention principle based on the notion of reciprocity, which means that when you work, you receive pay and benefits in return. Related research suggests that similar 'psychological contracts' develop with volunteers who also expect something from the organization in exchange for their labor (Blackman & Benson, 2010; Farmer & Fedor, 1999; Walker et al., 2016). In contrast to pay,

PHOTO 9.2
NVOAD awards National Volunteer of the Year to Paul Unruh, Mennonite Disaster Service.

reciprocity might include developing recognition programs and ways to affirm the value of a volunteer. Applying such findings to volunteers may be challenging, especially with SUVs who come and leave quickly. Nonetheless, it should be possible to launch some easy but heartfelt thanks as a way to recognize volunteers: through billboards, digital photos, social and traditional media, meals, and a simple thank you. Disaster organizations should be able to create more formal ways to thank volunteers from certificates to volunteer of the year awards (see Photo 9.2). In short, tell volunteers the organization needs them and values what they do (Devaney et al., 2015).

Conclusion

Volunteers make a difference. They do so for those who survive disasters and for the managers, community leaders, and officials who must guide their people through difficult and challenging recoveries. Volunteers also benefit from setting aside their own needs, work, and lives to go forth and take care of others harmed by terrible events. Involving a range of ages, cultures, and abilities can further disseminate the anticipated benefits of volunteering to more people. Survivors experience that volunteer effort as a means to lift their spirits and restore wholeness to their lives. Disaster volunteering thus represents more than picking up a hammer or a paintbrush, because the act of service generates both tangible and intangible results. Rebuilding or repairing a home matters, because it means people can return to normal, go back to work, bring their families home, and move on with their lives again. The repaired or rebuilt home also represents something more: an authentic connection between strangers, a healing moment in time, and a rebirth of the human spirit.

Emergency managers, volunteer coordinators, disaster organizations, and community leaders would be wise to build partnerships before a disaster to secure this rising of the human spirit for their people. Retaining those volunteers matters as well, in order to keep paying it forward for other disasters as well as for their own communities– to produce highly desired spillover effects. Retaining volunteers means preparing them to serve, providing an organized framework in which to do, and recognizing the value they bring in all their rich diversity. The effort is absolutely worth it to produce the benefits seen in this chapter.

References

Abramson, D., & Culp, D. (2013). *At the crossroads of long-term recovery: Joplin, Missouri six months after the May 22, 2011 tornado*. New York: National Center for Disaster Preparedness, Columbia University.

Adams, D. (1990). *Freedom summer*. NY: Oxford University Press.

Bachner, G., Seebauer, S., Pfurtscheller, C., & Brucker, A. (2016). Assessing the benefits of organized voluntary emergency services. *Disaster Prevention and Management, 25*(3), 298–313.

Barton, A. (1970). *Communities in disaster*. Garden City, NY: Doubleday and Company.

Benenson, J., & Stagg, A. (2016). An asset-based approach to volunteering: Exploring benefits for low-income volunteers. *Nonprofit and Voluntary Sector Quarterly, 45*(15), 131S–149S.

Becker, P., & Dhingra, P. (2001). Religious involvement and volunteering: Implications for civil society. *Sociology of Religion, 62/3*, 315–335.

Berg, A., & Johansen, O. (2017). Peeling potatoes as health promotion? Self-perceived benefits of volunteering among older adult volunteers in a Norwegian volunteer centre. *Open Journal of Social Sciences, 5*, 171–189.

Blackman, D., & Benson, A. (2010). The role of the psychological contract in managing research volunteer tourism. *Journal of Travel and Tourism Marketing, 27*(3), 221–235.

Bokszczanin, A. (2012). Social support provided by adolescents following a disaster and perceived social support, sense of community at school, and proactive coping. *Anxiety, Stress, & Coping, 25*(5), 575–592.

Branch-Smith, C., & Pooley, J. (2010). Women firefighters' experiences in the Western Australian volunteer bush fire service. *The Australian Journal of Emergency Management, 26*(3), 12–18.

Bruce, L. (2006). Count me in: People with a disability keen to volunteer. *Australian Journal on Volunteering, 11*(1), 59–64.

Chur-Hansen, A., Werner, L., McGuiness, C., & Hazel, S. (2015). The experience of being a guide dog puppy raiser volunteer: A longitudinal qualitative collective case study. *Animals: An Open Access Journal From MDPI, 5*, 1–12.

Corporation for National and Community Service. (2006). The power of help and hope after Katrina by the numbers: Volunteers in the Gulf. Available from https://www.nationalservice.gov/pdf/katrina_volunteers_respond.pdf, Accessed 23.07.19.

Devaney, C., Kearns, N., Fives, A., & Canavan, J. (2015). Recruiting and retaining older adult volunteers: Implications for practice. *Journal of Nonprofit & Public Sector Marketing, 27*, 331–350.

Erasmus, B., & Morey, P. (2016). Faith-based volunteer motivation. *Voluntas, 27*, 1343–1360.

Farmer, S., & Fedor, D. (1999). Voluntary participation and withdrawal: A psychological contract perspective on the role of expectations and organizational support. *Nonprofit Management and Leadership, 9*(4), 349–367.

Guiney, H., & Machado, L. (2018). Volunteering in the community: Potential benefits for cognitive aging. *Journals of Gerontology: Psychological Sciences, 73*(3), 399–408.

Hewstone, M., & Swart, H. (2011). Fifty-odd years of inter-group contact: From hypothesis to integrated theory. *British Journal of Psychology, 50*, 374–386.

Ho, H. (2017). Elderly volunteering and psychological well-being. *International Social Work, 60*, 4. Available from https://doi.org/10.1177%2F0020872815595111.

Isham, J., Kolodinsky, J., & Kimberly, G. (2006). The effects of volunteering for nonprofit organizations on social capital formation: Evidence from a statewide survey. *Nonprofit and Voluntary Sector Quarterly, 35*(3), 367–383.

Kako, M., & Ikeda, S. (2009). Volunteer experiences in community housing during the Great Hanshin-Awaji earthquake, Japan. *Nursing and Health Sciences, 11*, 357–359.

Kim, H. (2014). Learning from UK disaster exercises: Policy implications for effective emergency preparedness. *Disasters, 38*(4), 846–857.

Lum, T., & Lightfoot, E. (2005). Effects of volunteering on the physical and mental health of older people. *Research on Aging, 27*, 31–55.

Merchant, R., Leigh, J., & Lurie, N. (2010). Health care volunteers and disaster response – first, be prepared. *The New England Journal of Medicine, 362*(10), 872–873.

Nakhaei, M., Khankeh, H., Maoumi, G., Hosseini, M., Parsa-Yekta, Z., Kurland, L., & Castren, M. (2015). Impact of disaster on women in Iran and implication for emergency nurses volunteering to provide urgent humanitarian aid relief: A qualitative study. *Australasian Emergency Nursing Journal, 18*, 165–172.

National Council on Disability. (2009). *Effective emergency management: Making improvements for communities and people with disabilities*. Washington D.C: National Council on Disability.

Phillips, B. (2014). *Mennonite Disaster Service: Building a therapeutic community after the Gulf Coast storms*. Lanham, MD: Lexington Books.

Phillips, B. (2015). Therapeutic communities in the context of disaster. In A. Collins, S. Jones, B. Manyena, & J. Jayawickrama (Eds.), *Hazards, risks, and disasters in society* (pp. 353–371). Amsterdam: Elsevier.

Piliavin, J., & Siegl, E. (2007). Health benefits of volunteering in the Wisconsin longitudinal study. *Journal of Health and Social Behavior, 48*, 450–464.

Pine, N., Tarrant, R., Lyons, A., & Leathem, J. (2018). Teenagers' perceptions of volunteering following the 2010-2011 Canterbury earthquakes, New Zealand. *Journal of Loss and Trauma, 23* (5), 366–380.

Quevillon, R., Gray, B., Erickson, S., Gonzalez, E., & Jacobs, G. (2016). Helping the helpers: Assisting staff and volunteer workers before, during, and after disaster relief operations. *Journal of Clinical Psychology, 72*(12), 1348–1363.

Santini, Z., Meilstrup, C., Hinrichsen, C., Nielsen, L., Koyanagi, A., Krokstad, S., ... Koushede, V. (2019). Formal volunteer activity and psychological flourishing in Scandinavia: Findings from two cross-sectional rounds of the European social survey. *Social Currents, 6*(3), 255–269.

Seeley, S. (2014). Personal correspondence. With permission.

Shandra, C. (2017). Disability and social participation: The case of formal and informal volunteering. *Social Science Research, 68*, 195–213.

Sinclair, H., Doyle, E., Johnston, D., & Paton, D. (2012). Assessing emergency management training and exercises. *Disaster Prevention and Management, 21*(4), 507–521.

Stukas, A., Hoye, R., Nicholson, M., Brown, K., & Aisbett, L. (2016). Motivations to volunteer and their associations with volunteers' well-being. *Nonprofit and Voluntary Sector Quarterly, 45* (1), 112–132.

Taylor, A., & Frazer, A. (1981). The stress of post-disaster body handling and victim identification work. *Journal of Human Stress, 8*, 4–12.

Thoits, P., & Hewitt, L. (2001). Volunteer work and well-being. *Journal of Health and Social Behavior, 42/2*, 115–131.

Thompson, J. (1991). The management of body recovery after disasters. *Disaster Management, 3* (4), 206–210.

Thormar, S., Gersons, B., Juen, B., Maschang, A., Djakababa, M., & Olff, M. (2010). The mental health impact of volunteering in a disaster setting: A review. *The Journal of Nervous and Mental Disease, 198*(8), 529–538.

Unruh, P. (2017). *Long journey home: Stories of hope and healing*. Lititz, PA: Mennonite Disaster Service.

van Willigen, M., Edwards, B., Edwards, T., & Hessee, S. (2002). Riding out the Storm: The Experiences of the Physically Disabled during Hurricanes Bonnie, Dennis, and Floyd. *Natural Hazards Review, 3/3*, 98–106.

Walker, A., Accadia, R., & Costa, B. (2016). Volunteer retention: The importance of organizational support and psychological contract breach. *Journal of Community Psychology, 44*(8), 1059–1069.

Wang, X., Ip, P., & Chan, C. (2016). Suicide prevention for local public and volunteer relief workers in disaster-affected areas. *Journal of Public Health Management Practice, 22*(3), E39–E46.

Whittaker, J., McLennan, B., & Handmer, J. (2015). A review of informal volunteerism in emergencies and disasters: Definition, opportunities and challenges. *International Journal of Disaster Risk Reduction, 13*, 358–368.

Wilson, J., & Musick, M. (1997). Who cares? Toward an integrated theory of volunteer work. *American Sociological Review, 62*(5), 694–713.

Further reading

Farris, A. 2006. *Katrina anniversary finds faith-based groups still on the front lines, resilient but fatigued*. http://www.religionandsocialpolicy.org. Accessed 15.01.08.

Hua-Chin, H. (2017). Elderly volunteering and psychological well-being. *International Social Work, 60*(4), 1028–1038.

CHAPTER 10

The future of disaster volunteerism

The future of disasters and disaster volunteering

Disaster scientist E.L. Quarantelli noted in 1991 that we would experience bigger and newer disasters. Dr. Quarantelli was right. Since then, the world has seen massive events that have caused catastrophic damage and loss of life: the 2004 tsunami that impacted 13 nations and killed over 300,000 people; the 2010 Haiti earthquake that devastated its capital city region and claimed over 200,000 souls; and the cascading Japanese event that began as an earthquake, caused a tsunami, undermined the integrity of a nuclear plant, and took over 20,000 lives in multiple cities. Such massive loss of life and catastrophic damage is stunning. Even more alarming losses have continued to escalate in recent decades, suggesting that the world will face even graver concerns in the future. Disaster volunteers will be needed more than ever (McLennan, Whittaker, & Handmer, 2016).

Data collected by the United Nations Centre for Research on the Epidemiology of Disasters (CRED, 2018) focused on a time period from 1998 to 2017 and compared the data to the prior 20 years. CRED empirically confirmed Quarantelli's findings:

- Economic losses increased 68% from the prior time period of 1978–1991.
- Losses from extreme weather increased the most, by 151%.

In addition:

- Over 90% of the disasters came from floods, storms, and extreme weather related to climate change including droughts and heatwaves.
- 1.3 million people died.
- 4.4 billion people needed help with injuries, loss of homes, emergency food, relief supplies and recovery.
- Disaster damages cost an average of $520 billion per year in U.S. dollars.

Disaster Volunteers. DOI: https://doi.org/10.1016/B978-0-12-813846-5.00010-1

Newer events that qualify as disasters, defined as events that disrupt community functioning (for a robust discussion, see Perry & Quarantelli, 2005; Quarantelli, 1998), have emerged as well: the terrorism of September 11 and attacks on embassies, tourist sites, places of worship, malls, and outdoor concerts; cyberattacks that include ransomware holding health care and law enforcement agencies hostage; geomagnetic storms with the potential to undermine the power grid and cause loss of life; and climate change that increases wildfires and produces heatwaves, and threatens islands, coastal regions, people, and cultural heritage. New ways to volunteer have also appeared, including through a variety of technological resources, stretching the distances through which people can serve. Disaster volunteers are needed more than ever, and in ways that we never anticipated.

Why are disasters increasing and even worsening? Experts tell us it is because we continue to live in areas of significant risk near coastlines, earthquake faults, and hazardous materials sites (Quarantelli, 1991). Given population growth, and people locating in areas where work provides opportunities to earn income and feed one's family, disaster risks will continue. Coastal areas, for example, will become increasingly subject to extreme weather such as hurricanes or cyclones that generate flooding. Given climate change, coastal and island areas will continue to experience heightened risks that threaten critical elements in societies like infrastructure in the U.S. (Neumann et al., 2015), tourist economies in South Africa (Fitchett, Grant, & Hoogendoorn, 2016) and food and health in the Mediterranean and on tropical islands (Cramer et al., 2018; Hernández-Delgado, 2015). Without mitigative actions, humanitarian help will be needed.

Continuing and increasing hazards means that disaster volunteers will still be called upon, including those able to face new challenges and in new ways that provide help. In what ways might people volunteer to meet these new disasters? Consider these possibilities:

- *Terror attacks*. Given the prevalence and increases in terror attacks worldwide, volunteer effort will be needed. Terror attacks produce structural damage, and particularly seek to cause psychological damage and fear. Consequently, traditional volunteer opportunities might not exist. In Nairobi, people responding to the 2013 Westgate Mall attack opened feeding stations, provided spiritual and emotional comfort, raised funds for people with injuries or for burials, and promoted a strong sense of community resilience. Musicians held fundraising and solidarity concerts after September 11. Online donation drives raised money for victims of the 2015 Paris concert attacks. When people sustain injuries, they may also need assistive or prosthetic devices which volunteers could fundraise around or secure through donations.

Quite often, such devices are specialized to the individual, so crowdsourcing funds represents a best way forward to buy just the right device. In some instances, securing walkers, wheelchairs, or similar devices can suffice – but, always ask first to make sure that donations are appropriate. Volunteers could also offer transportation to people injured and in need of help getting to physical and occupational therapy or provide child care so they can travel alone. Other possibilities could be followed as well, such as long distance runs that provide fitness benefits, bring people together, and raise money – like the Oklahoma City Memorial Marathon that raises funds for education and healing around acts of terrorism. New events also yield unexpected opportunities to help, such as the thousands of strangers who attended a funeral after a massive hate crime attack in El Paso, Texas in 2019. People came from hundreds of miles away and waited hours in 100 Fahrenheit degree heat to console her husband.

- *Cybercrimes as disasters*. Next, consider how Internet-based crimes can disrupt community functioning. Cyberattacks, from criminal behavior to governmental warfare, can interrupt a hospital held through ransomware or an election system compromised by interference from another government. Volunteers could have a role in educating people about cybercrimes (what emergency managers call preparedness) or encouraging compliance with anti-phishing campaigns in the workplace (what emergency managers call mitigation). An attack on the electrical power grid will prove deadly in winter, similar to what could occur with geomagnetic storms. Help will be needed with shelters and mass care.
- *Geomagnetic storms*. It may seem like solar storms occur far away and perhaps without direct impact. After all, they are most likely to disrupt cell phones or navigational devices which may not seem like they apply to the volunteering public. But consider the impact of such storms: inabilities to communicate, problems reaching children or elderly via phones, and the potential impact if a storm takes out a power grid. How will we care for people without power – who need oxygen, assistive devices, surgeries, air conditioning or heat? In most disasters, we open shelters which may be the case if another Carrington event occurs. How will we transport people to shelters, including senior citizens and people with medical conditions? How long would they be there, needing food, hydration, showers, medical care, and social interaction? We will need volunteers in such places. What about place or home-bound people - is it unrealistic to think of volunteers serving on generator teams to keep people alive or providing a foster care-like situation for people, pets, and livestock?

- *Pandemics*. Globally, a significant amount of effort has gone into pandemic planning. The potential exists for a virus to spread quickly due to international air travel – and has. Increasingly, we are seeing communities prepare for pandemics by recruiting volunteers into helping with Points of Distribution (PODs). In such locations, medications would be passed out in order to reduce the spread of infectious disease. Volunteers play a role in helping with traffic, filling out paperwork, routing people through the POD, and acting as qualified health care volunteers. International volunteers will be especially needed. The repeated outbreaks of Ebola, the challenges presented by H1N1 influenza (Kuster et al., 2013), and other diseases have shown that newer forms of infectious diseases will require attention to reduce loss of life and a pandemic.
- *Climate change*. The realities of climate change are upon us, including extreme heat, cold, drought, and wildfire. One study in Vietnam found that people's exposure to riverine and coastal flooding would increase exposure from one-third of the population to about 46% depending on the severity of the flood (Bangalore, Smith, & Veldkamp, 2019). In addition to people at risk, so are the flora and fauna around us and the buildings in which we live and work. Straightforward volunteer acts like opening warming and cooling centers for people at risk represent one volunteer option given the debilitating and deadly consequences of heatwaves, polar extremes, and droughts. Consider also how we protect the outdoors, such as the Native Animal Network of Southern Australia's volunteer efforts to save wildlife from bushfires. Coastal zones also bear disproportionate risk of losing cultural and historic heritage, which volunteers can help to protect through mitigation work or by helping area conservators and historic preservationists to raise awareness and funds. The United Nations Educational, Scientific, and Cultural Organization (UNESCO) has addressed such climate changes for world heritage sites. The Chan Chan, Peru site, for example, has experienced erosion from rainfall resulting in damage to the earthen structure bases (UNESCO, 2010). Volunteer efforts might look different for such disasters, and involve help in supporting preservationists' efforts, advocating to save world heritage sites, and fundraising to mitigate damage.

Another global condition that will inspire continued involvement in disaster volunteering is the reality that significant income inequality exists. Whether within or between a nation, income influences the degree of vulnerability that a household may experience as well as the resources they will have to be more resilient. People may not be able to secure housing that affords them protection from the elements, or they may need to live in areas of high-hazard or repetitive risks due to their livelihoods or income levels. The CRED (2018) study that started this chapter found that disasters truly harm people in lower income countries the most. From 1998–2017, disasters pushed 26 million people into poverty. Where one lives may matter as well. Living in lower income countries disasters deepen poverty and inequality as found in Vietnam (Bui, Dungey, Viet Nguyen, & Phuong Pham, 2014), Mexico (Rodriguez-Oreggia, De La Fuente, De La Torre, & Moreno, 2013), and Sri Lanka (Keerthiratne & Tol, 2018). In one study, cross-national income inequality increased over what must have been a difficult a five-year period post-disaster then finally began to decrease (Yamamura, 2015).

Income inequality also affects different populations in rather adverse ways. Coupled with historic patterns of discrimination, such as not hiring people with disabilities, means that as a group they remain under-employed, under-paid, and at risk when disasters strike. Seniors also lose income over time as they retire or become pensioners. Add in health care challenges, and savings dwindle. Single parent households remain significantly below the poverty levels worldwide. In short, disaster volunteers will always be needed because of the challenges that income inequality presents.

Where volunteers will be needed

In Chapter 5, readers learned about the life cycle of disasters in which emergency managers organize their activities: mitigation (reducing impacts), preparedness and planning (getting ready), response (acting to save lives and property), and recovery (working to pick up the pieces and enable people to return to homes and work). Historically, volunteers have been drawn to the response and recovery time periods. Response calls to people because they see compelling images and feel an obligation to provide help and relief. Recovery, when the majority of volunteers are needed the most, also brings in many experienced organizations and volunteers to help. But the four phases could all use volunteers. Consider these possibilities:

- *Mitigation*. In an ideal world, we would mitigate disasters in order to be as resilient as possible. That is not the reality, however, as people continue to move to hazardous areas for their livelihoods. In some places, people cannot relocate like in Mozambique, Malawai and Zimbabwe where heavy floods killed over 1000 in 2019. What can volunteers do to mitigate risk? Efforts can be put into place to help people evacuate and to swim as a last resort, an effort that NGO volunteers facilitated in Sri Lanka and India after the 2004 tsunami. Volunteers can also serve on mitigation planning committees in their communities, and participate in assessing hazards, identifying and prioritizing ways to be more resilient, and helping local government convince the public to do something before disasters occur (for resources, see https://www.fema.gov/local-mitigation-planning-resources). Mitigation work also fits well with governmental efforts to promote community based disaster risk reduction and building community resilience (Chandra et al., 2013; McLennan et al., 2016). Volunteers can become involved in measuring and installing plywood shutters in hurricane areas (Federal Alliance for Safe Homes, 2013) or assist people with earthquake reduction measures like securing bookcases or cupboards. Volunteers can also help homeowners reduce highly flammable landscaping around their homes and put in plants resistant to wildfire (Porensky, Perryman, Williamson, Madsen, & Leger, 2018).

- *Preparedness*. The preparedness phase represents a wonderful opportunity for volunteer participation. After all, if volunteers help people get ready for a disaster it might make everyone's life safer coupled with the ability to rebound more effectively. What if volunteers helped to educate people in workplaces, schools, civic clubs, high rise apartment buildings, transportation sites, parks, sports venues, airports, and concert settings? Would that extra layer of alertness caused by education decrease losses associated with active aggressors or during a severe storm at a large outdoor event? It very well could, in an activity easily taken on by scouting troupes, civic clubs, and neighborhood associations. Collective volunteer efforts could focus on high risk populations like senior citizens, people with disabilities, children, pets, and livestock as well as protecting cultural and historic sites at risk from climate change-related flooding or terror attacks.
- *Planning*. Everyone needs to be at various planning tables around disasters from individual household planning to business planning, as well as community or even national and international planning efforts. The reason is because we all offer useful insights helpful to those who work to keep us safe. For example, emergency managers can benefit from the perspectives of disability advocates who understand specific needs and offer valuable insights - and connections to people who want to volunteer. Involvement of low-income and elderly residents in evacuation planning might have saved hundreds of lives when Katrina tore through New Orleans. Or, something as straightforward as involving women in recovery planning would have eliminated post-tsunami recovery problems such exclusion created in India. How can volunteers participate? Readers can start in their own family by creating communication and evacuation plans for area hazards. Volunteers can convene their own neighborhood in a planning effort, from something as straightforward as how to be ready for minor street flooding to search and rescue after a tornado. Tools exist for this (see Box 10.1).

Box 10.1 Planning tools

For personal and workplace communication and hazard-specific plans, visit www.ready.gov in the U.S. and https://www.civildefence.govt.nz/ in New Zealand.

At the neighborhood level, consider a program called Map Your Neighborhood to pre-identify areas of risk and resources to use for various hazards. For an example, visit https://mil.wa.gov/map-your-neighborhood.

How do you keep your business functioning after a disaster? What would you do after a geomagnetic storm or cyberattack let alone area flooding or high winds? Business continuity planning templates and tools can be found at the FEMA library, https://www.fema.gov/media-library/assets/documents/89510.

Are you at a university or college? Does your institution have a plan to survive a disaster or terror event? For a useful set of disaster resistant university tools and approaches, go to https://www.fema.gov/media-library/assets/documents/2288.

Volunteers can also join a Local Emergency Planning Committee or an area safety council, volunteer to help develop an evacuation plan, or join the effort to plan for and carry out a pandemic planning initiative. If such opportunities do not exist locally, consider creating an emergent organization to address hazards and risks.

- *Response*. Volunteers will continue to be needed during the response phases, especially if they prepare ahead of time and work in concert with responding officials and affiliate with experienced disaster organizations. Increasingly, emergency managers have been moving to both discourage and manage spontaneous volunteers. This kind of carefully coordinated and managed effort will be essential as disasters worsen and impact more people, especially those at lower income levels. It is likely that personal and electronic convergence will continue into the foreseeable future. It may be necessary to offer newer kinds of training considering the onset of so many active attacks, like first aid and CPR training as well as the best ways to move and transport a critically injured individual. How might volunteers take action during a cyberattack? Could digital volunteers play a role in the future?
- *Recovery*. Similar to response, volunteers will always be needed during the recovery time period, especially those with an array of skills from the human to the built environment and including affected flora and fauna. Rebuilding lives and communities takes time, resources, and dedication. Emergency managers, volunteer coordinators, and disaster organizations will need to continue to be involved in recruiting and retaining volunteers and certainly at higher levels than in previous years. Fresh perspectives will be required, including volunteers committed to rebuilding homes in ways that prove resilient. Green rebuilding will also increase during disasters, in an effort to promote

environmentally responsible efforts. As technology increases, smart rebuilding may increase requiring input and expertise from digital volunteers. Accessibility and universal design will also continue, involving volunteers in such efforts as well as disability advocates and people with disabilities in mutual efforts to rebuild homes in which people can live well.

As the CRED report noted, the last twenty years have seen worsening impacts from disasters. The next twenty years will continue that trend unless significant interventions take place, and will see newer ways in which people and places experience harm. But the volunteer remains promising for an even broader and deeper set of volunteers to participate.

Where volunteers will come from

Another anticipated change in the future of volunteering is the source of such efforts. Where will volunteers come from? A scan of relevant studies and reports indicate that the future looks bright for both traditional and new sources of disaster volunteers. Several promising sources are likely to yield new volunteers: digital help, the private sector, and local community-based efforts. In addition, emergency managers and volunteer coordinators can continue to rely on traditional sources of help, particularly from the faith-based sector. NVOAD type efforts are becoming increasingly coordinated and effective including integrating new organizations into this volunteer family. The United Nations continues to integrate volunteers and voluntary organizations into its numerous efforts along with detailed guidance on volunteering policies (see https://www.unocha.org/). NGO and INGO's will continue to need volunteers to extend their reach into underserved areas affected by disasters. Sources exist, if managers take advantage of working with them and their volunteers and if volunteers connect into the network of disaster organizations already experienced and ready to help.

Traditional sources of disaster volunteers

Given that the majority of disaster organizations currently come from a well-established and diverse faith-based community, this traditional base of disaster volunteers will continue. Managers and organizations will want to continue to rely on the social networks and inspirational messaging found in these locations to recruit volunteers. Civic clubs also represent traditional places to secure help and funding for volunteer efforts. With many such clubs located worldwide, as well as the numerous affiliated units that connect to civic interests, this traditional voluntary base should continue to produce recruits.

A key demographic for future volunteers will come from age groups, with retirees representing a particularly rich source for traditional volunteering. The most likely pool of retirees seems to be those who have fairly recently retired and who are looking for a way to continue meaningful work. Managers and organizations would be wise to develop ties to retiree groups, housing sites, and communities. Recent retirees will also probably be in the best health as well, and volunteering can be offered as a benefit to continue their well-being. But even older volunteers should not be overlooked, as they may also want to travel to disaster sites or support efforts from their home base.

Younger volunteers may be more challenging to attract, especially as they move into the workforce and start families. But by working with corporate policies that inspire volunteering, or encouraging local businesses to start such efforts, managers and organizations may be able to diversity the age demographic profile of their volunteers. It would be wise to develop intentional programs that enable participation specific to age or as a way to participate locally. Organizations that build home components at a site distant from the disaster and then transport those sections offer such an option. That is how the Partnership Home Project works, by building and then trucking sections of a home to a needed site (see https://mds.mennonite.net/partnership-home-project/). Traditional sources of volunteers can be relied on, and increasingly so as managers make it possible to do so through means that enable participation. What if such an effort took place on corporate grounds? For younger aged volunteering, managers may need to take the work to them or they may need to ask their workplaces to make it so.

What is equally clear is that managers and organizations have to make an effort to ask people to volunteer. Asking is particularly important to encourage a diverse set of volunteers to appear, including people from varying racial and ethnic groups, income levels, and abilities. Managers and voluntary organizations will need to get out of their offices and go to where people interact in their social networks to make that ask, and to involve trusted members of those networks to facilitate the invitation.

Newer sources of disaster volunteers

Not surprisingly, electronic convergence will represent one newer stream of help, from those who spontaneously appear and those using digital tools to volunteer (McLennan et al., 2016). Emergency managers have at times been reluctant to move toward such digital volunteerism because of both volume and accuracy problems with social media. However, by involving emergency management practitioners (including those who volunteer) with the public, digital volunteers can assess the content and counteract incorrect information (Roth & Prior, 2019). As one example, Virtual Operations Support Teams (VOSTs) emerged as a hybrid effort from emergency management staff who volunteered to monitor digital communications (Roth & Prior, 2019). A study of VOSTs in the U.S., Scotland, Germany, France, Spain, and Canada found they provided value despite the volume of social media communications that produced both actionable content and total hoaxes (Roth & Prior, 2019). In addition, VOSTs tend to be connected to each other, thus providing a means to leverage their networks in a given event (Roth & Prior, 2019). However, VOSTs and digital volunteer organizations remain yet to be well developed or integrated into emergency management operations (McLennan, Whittaker, & Handmer, 2015) but may be cresting a volunteer wave of the future (for an example, see http://vostvic.net.au/post/virtual-operations-support-team-vost-victoria).

Another anticipated source of volunteer help comes from the private sector (McLennan et al., 2016) which has the potential to offer considerable skill sets for disaster efforts. As a consequence, the future of public-private partnerships appears quite promising considering trends toward corporate philanthropy that also emphasize community engagement. In a study in the U. S., a survey for the Points of Light Foundation discovered that 82% of the fifty most civic-engaged corporations said that volunteering boosts employee morale and productivity (Points of Light, 2014). The bulk of the top fifty companies also offered paid time off to volunteer. Research supports similar findings, suggesting that when companies promote corporate social responsibility, the broader community benefits and customer loyalty increases (Korschun, Bhattacharya, & Swain, 2014). Australian researchers report that while connections between corporations and emergency volunteering tend to be short-lived and response-oriented, opportunities exist to improve opportunities (McLennan et al., 2015). Managers and organizations thus need to make more of corporate interests in disaster contexts.

Traditionally, we think of disaster volunteers as leaving their work and homes to go where an event has occurred. But disasters are inherently local events and affect our own communities. A move toward community-based disaster risk reduction means that we must increasingly invite and involve local volunteers in efforts that reduce the impacts of disasters (Maskrey, 2011; McLennan et al., 2016). Many opportunities exist, yet public agencies from emergency management and beyond often fail to do so (Stajura et al., 2012). When we lack local volunteer involvement in disaster reduction work, we fail our own communities and limit our abilities to fully engage those at highest risk (Berke, Cooper, Salvesen, Spurlock, & Rausch, 2011; Stajura et al., 2012). Such local involvement lies at the heart of multiple international frameworks to reduce disaster impacts (see Box 10.2).

Box 10.2 International frameworks and local participation

International frameworks have increasingly called for community involvement at the local levels. Volunteer involvement is key to initiating these frameworks, which are designed to make the world a safer place:

- *Hyogo Framework for Action*, 2005–2015. The Hyogo effort encouraged making disaster preparedness a national and local priority by promoting a culture of safety (see https://www.unisdr.org/we/coordinate/hfa). Community participation lies at the heart of several of the five priorities, and the last specifically calls for voluntarism and participation.
- *Sendai Framework*, 2015–2030. Similarly, the Sendai Framework calls for action to reduce losses and build more resilient communities which will require public and private partnerships and participation. Sendai specifically calls for women and people with disabilities to "publicly lead and promote gender-equitable and universally accessible approaches during the response and reconstruction phases." (See https://www.unisdr.org/we/inform/publications/44983).
- *Global Platform for Disaster Risk Reduction*, 2017. The Global Platform is updated biannually at an international conference where high level officials interact with people at local levels. Their collective effort is designed to implement efforts like the Sendai Framework and to continue commitment to reducing disasters. At the Platform conference in Cancun, Mexico in 2017, volunteers in grass roots organizations presented their efforts to save their homes and communities. Even in places with limited resources, grass roots volunteers accomplished a great deal.

Conclusion

Traditionally, many books and articles in disaster science end with implications for the policies, practices, and research needed yet to improve a particular challenge. The good news about disaster volunteers is that people always seem willing to serve. To make the most of these eager volunteers, the last section of this volume lists action steps that can improve the knowledge base and practices around disaster volunteerism.

Policy

What kinds of policies should inform volunteering and the management of helping? Some possible places for advancement in policies on volunteering include:

- Reconsidering Good Samaritan laws for their reach, impact, and gaps related to the protection of volunteers and volunteer organization.
- Reviewing policies that pertain to safety concerns and work sites for volunteers in order to provide consistent messaging and standards.
- Developing workplace policies that encourage volunteering including a means to receive paid time off to volunteer.
- Requiring volunteer management plans at local, regional, national and international levels that embrace local demographics.

- Embracing best practices such as careful use and management of spontaneous volunteers.
- Mandating rainy day funds to support volunteer integration into emergency management efforts.

Practice

A study from Australian emergency managers where over 250,000 volunteers serve in fire, ambulance, and emergency services found that emergency managers need support to leverage those volunteers (Kruger & McLennan, 2018). The kind of support the emergency managers needed required local government help to facilitate a culture of local engagement. Local leaders can set expectations about such civic involvement, inspire participation, and provide funding to make it happen. That work can arise out of a range of interconnections between government (including emergency managers), nongovernment organizations (including disaster organizations), traditional and new kinds of volunteers, and workplaces. Best practices for future volunteers would thus seem to require:

- Establishing working relationships between emergency managers and voluntary agencies. Get to know each other.
- Writing volunteer management plans between emergency managers and volunteer coordinators.
- Testing and revising volunteer management plans.
- Pre-establishing Volunteer Reception Centers and online registries.
- Creating memoranda of agreement between emergency managers, volunteer coordinators, voluntary agencies and related units.
- Offering opportunities for volunteers to stay active in events and activities related to emergency management in order to retain interest, promote service, and offer a chance to use skills.
- Developing an electronic means to manage volunteers and link to area needs when a disaster occurs.
- Making a way for digital volunteers to become involved from a distance.
- Supporting each other's recruiting efforts.
- Following the principles for retention of volunteers.
- Setting aside funds to support volunteer management for recruitment, training, and retention.
- Supporting initiatives, programs, and policies that work toward reducing risks from the local through international levels.

Research

In researching scholarly works for this volume, it became clear that a significant push needs to occur to build the body of knowledge on disaster volunteering.

Doing so will inform the policies and practices around such help, in order to produce the best possible outcomes. Disaster science needs to embrace volunteering further as a research topic that should be viewed as a significant gap in our knowledge base that is essential to fill. Scholars working in this area might benefit from:

- Producing case studies of organizations that surface evidence-based best practices throughout the lifecycle of disasters: mitigation, preparedness, planning, readiness, response, and recovery.
- Generating comparative case study analysis of organizations by type and location. This will require producing case studies for use in a comparative cross-case investigation as indicated in the first bullet.
- Determining what benefits accrue specifically to disaster volunteers and if those findings conform to the findings from studies of volunteering in general.
- Confirming the most effective ways to recruit and retain volunteers for disaster service.
- Revealing, systematically, the best practices for international volunteering within and across cultural and linguistic differences as well as within various social, political, and economic contexts.
- Determining a set of standards for training both spontaneous and affiliated volunteers. How should managers prepare spontaneous volunteers to conduct immediate work? What are the best ways to socialize affiliated volunteers for long-term volunteer retention?
- Surveying volunteers for their experiences on the full extent of their involvement from recruitment through training, work site efforts, recognitions and rewards, debriefing, and retention.
- Studying the best ways to debrief volunteers, particularly those who may experience exposure to potentially traumatic disaster experiences.
- Interviewing emergency managers and volunteer coordinators to find out the best ways to manage volunteers.
- Conducting a stepwise assessment of volunteer management plans to find commonalities, identify missing areas that should be addressed, and best practices.
- Observing and analyzing tabletop and field exercises that test volunteer management plans.
- Conducting on-site studies of volunteer management plans in actual disasters: what works and what does not work, and why?
- Studying safety training, injuries, and deaths of volunteers at disaster sites.
- Engaging in policy analysis of the impacts of various policies, including Good Samaritan laws.
- Investigating liability concerns related to volunteers and volunteer management.

Regardless of where you are located, as a volunteer, an emergency manager or volunteer coordinator, the leader of an organization or a workplace – you can be part of the future of disaster volunteering. Disasters will continue to happen, including ones that we have not yet conceived of. People will need help and regardless of where you are located, you can be part of filling that need especially if we work together following the best practices outlined in this volume.

References

Bangalore, M., Smith, A., & Veldkamp, T. (2019). Exposure to floods, climate change, and poverty in Vietnam. *Economics of Disasters and Climate Change, 3*, 79–99.

Berke, P., Cooper, J., Salvesen, D., Spurlock, D., & Rausch, C. (2011). Building capacity for disaster resiliency in six disadvantaged communities. *Sustainability, 3*, 1–20.

Bui, A., Dungey, M., Viet Nguyen, C., & Phuong Pham, T. (2014). The impact of natural disasters on household income, expenditure, poverty and inequality: Evidence from Vietnam. *Applied Economics, 46*(15), 1751–1766.

Centre for Research on the Epidemiology of Disasters (CRED). (2018). *Economic losses, poverty, and disasters 1998-2017*. United Nations Office for Disaster Risk Reduction.

Chandra, A., Williams, M., Plough, A., Stayton, A., Wells, K., Horta, M., & Tang, J. (2013). Getting actionable about community resilience: The Los Angeles County community disaster resilience project. *American Journal of Public Health, 103*(7), 1181–1189.

Cramer, W., Guiot, J., Fader, M., Garrabou, J., Gattuso, J.-P., Iglesias, A., ... Xoplaki, E. (2018). Climate change and interconnected risks to sustainable development in the Mediterranean. *Nature Climate Change*, Nature Publishing Group, *8*(11), 972–980.

Federal Alliance for Safe Homes. (2013). Mitigation: An approach through volunteering. Available from http://www.nvoad.org/wp-content/uploads/2014/04/5-2-13_voad_session_presentation_-_no_videos_tim_smail.pdf, Accessed 29.07.19.

Fitchett, J., Grant, B., & Hoogendoorn, G. (2016). Climate change threats to two low-lying South African coastal towns. *South African Journal of Science, 112*(5/6), 1–9.

Hernández-Delgado, E. (2015). The emerging threats of climate change on tropical coastal ecosystem service, public health, local economies and livelihood sustainability of small islands. *Marine Pollution Bulletin, 101*, 5–28.

Keerthiratne, S., & Tol, R. (2018). Impact of natural disasters on income inequality in Sri Lanka. *World Development, 105*, 217–230.

Korschun, D., Bhattacharya, C., & Swain, S. (2014). Corporate social responsibility, customer orientation, and the job performance of frontline employees. *Journal of Marketing, 78*, 20–37.

Kuster, S. P., Coleman, B. L., Raboud, J., McNeil, S., De Serres, G., Gubbay, J., & McGeer, A. J. (2013). Risk factors for influenza among health care workers during 2009 Pandemic, Toronto, Ontario, Canada. *Emerging Infectious Diseases, 19*(4), 606–615. Available from https://dx.doi.org/10.3201/eid1904.111812.

Kruger, T., & McLennan, B. (2018). *Emergency volunteering 2030: Views from local government managers*. Melbourne: Bushfire and Natural Hazards CRC.

Maskrey, A. (2011). Revisiting community-based disaster risk management. *Environmental Hazards, 10*(1), 42–52.

McLennan, B., Whittaker, J., & Handmer, J. (2015). *Emergency volunteering in Australia: Transforming not declining*. Melbourne: Bushfire and Natural Hazards CRC.

McLennan, B., Whittaker, J., & Handmer, J. (2016). The changing landscape of disaster volunteering: Opportunities, responses and gaps in Australia. *Natural Hazards, 84*, 2031–2048.

Neumann, J., et al. (2015). Climate change risks to US infrastructure. *Climatic Change, 131*, 97–109.

Perry, R., & Quaratelli, E. L. (Eds.), (2005). *What is a disaster? New answers to old questions*. Xlibris, International Research Committee on Disasters.

Points of Light. (2014). *The Civic 50: A roadmap for corporate community engagement in America*. Washington D.C.: Points of Light Foundation.

Porensky, L., Perryman, B., Williamson, M., Madsen, M., & Leger, E. (2018). Combining active restoration and targeted grazing to establish native plants and reduce fuel loads in invaded ecosystems. *Ecology and Evolution, 8*, 12533–12546.

Quarantelli, E. L. (1991). *More and worse disasters in the future*. Newark, DE: Disaster Research Center, Preliminary Paper #158.

Quarantelli, E. L. (Ed.), (1998). *What is a disaster?? Perspectives on the question*. NY: Routledge.

Rodriguez-Oreggia, E., De La Fuente, A., De La Torre, R., & Moreno, H. (2013). Natural disasters, human development and poverty at the municipal level in Mexico. *Journal of Development Studies, 49*(3), 442–455.

Roth, F., & Prior, T. (2019). Utility of Virtual Operation Support Teams: An international survey. *Australian Journal of Emergency Management, 34*(2), 53–59.

Stajura, M., Glik, D., Eisenman, D., Prelip, M., Martel, A., & Sammartinova, J. (2012). Perspectives of community and faith-based organizations about partnering with local health departments for disasters. *International Journal of Environmental Research in Public Health, 2*, 2293–2311.

UNESCO. (2010). *Managing disaster risks for world heritage*. Rome, Italy: United Nations Educational, Scientific and Cultural Organization.

Yamamura, E. (2015). The impact of natural disasters on income inequality: Analysis using panel data during the period 1970 to 2004. *International Economic Journal, 29*(3), 359–374.

Index

Note: Page numbers followed by "*f*," "*t*," and "*b*" refer to figures, tables, and boxes, respectively.

U

V

W

Z

Printed in the United States
By Bookmasters